Lecture Notes in Computer Scie

Commenced Publication in 1973
Founding and Former Series Editors:
Gerhard Goos, Juris Hartmanis, and Jan van Leeuwen

Carlos Alberto Maziero   João Gabriel Silva
Aline Maria Santos Andrade
Flávio Morais de Assis Silva (Eds.)

Dependable Computing

Second Latin-American Symposium, LADC 2005
Salvador, Brazil, October 25-28, 2005
Proceedings

 Springer

Volume Editors

Carlos Alberto Maziero
Pontifícia Universidade Católica do Paraná
Programa de Pós-Gradução em Informática Aplicada
80.215-901 Curitiba PR, Brazil
E-mail: maziero@ppgia.pucpr.br

João Gabriel Silva
Universidade de Coimbra
Dep. Eng. Informatica - Polo II
Pinhal de Marrocos, 3030-290 Coimbra, Portugal
E-mail: jgabriel@dei.uc.pt

Aline Maria Santos Andrade
Flávio Morais de Assis Silva
Universidade Federal da Bahia (UFBA)
Departamento de Ciência da Computação (DCC)
Laboratório de Sistemas Distribuídos (LaSiD)
Campus de Ondina - Prédio do CPD, Av. Adhemar de Barros, S/N, CEP 40170-110,
Salvador-BA, Brazil
E-mail: {aline,fassis}@ufba.br

Library of Congress Control Number: 2005933898

CR Subject Classification (1998): C.3, C.4, B.1.3, B.2.3, B.3.4, B.4.5, D.2.4, D.2.8, D.4.5, E.4, J.7

ISSN   0302-9743
ISBN-10   3-540-29572-0 Springer Berlin Heidelberg New York
ISBN-13   978-3-540-29572-3 Springer Berlin Heidelberg New York

Springer is a part of Springer Science+Business Media

springeronline.com

© Springer-Verlag Berlin Heidelberg 2005
Printed in Germany

Typesetting: Camera-ready by author, data conversion by Scientific Publishing Services, Chennai, India
Printed on acid-free paper   SPIN: 11572329   06/3142   5 4 3 2 1 0

Foreword

The *Latin-American Symposium on Dependable Computing*, LADC, is the main Latin-American event dedicated to the discussion of the many issues related to dependability in computer systems and networks. It is a forum for researchers and practitioners (from all over the world) to present and discuss their latest results and experiences in this field.

LADC 2005, the second edition of this event, followed on the success of LADC 2003, which took place at the Polytechnic School of the University of São Paulo. It was organized by LaSiD, the Distributed Systems Laboratory associated with the Department of Computer Science of the Federal University of Bahia. LADC 2005 was sponsored by SBC, the Brazilian Computer Society, in cooperation with IEEE TC on Fault-Tolerant Computing, IFIP Working Group 10.4 "Dependable Computing and Fault-Tolerance", SADIO, the Argentine Society for Informatics and Operations Research, SCCC, the Chilean Computer Science Society, and SMCC, the Mexican Society for Computer Science.

LADC 2005 was structured around technical sessions, keynote speeches and a panel. Two workshops were co-located with LADC 2005: WDAS (Latin-American Workshop on Dependable Automation Systems) and WTD (3rd Workshop on Theses and Dissertations on Dependable Computing). WDAS is a forum where members of academia and industry can meet to discuss specific dependability issues related to automation systems. WTD is a student forum dedicated to the discussion of ongoing and recent work in the field of dependability carried out at graduate level.

We would like to thank the LADC 2005 Organizing Committee and the support staff at LaSiD for having helped us with the organizational tasks, the Steering Committee for their advice, and the chairs of the technical committees for their cooperation. A special "thank you" goes to Raimundo Macêdo and to Rogério de Lemos who were sources of constant support and suggestions. We are also grateful to Raimundo Macêdo for having suggested our names to chair this symposium. Additionally, we would like to thank the invited guests, all the authors of submitted papers, the support provided by CAPES (Coordenação de Aperfeiçoamento de Pessoal de Nível Superior), the sponsoring partners, and Springer for accepting to publish the LADC proceedings in the LNCS series.

We hope all present at LADC 2005 enjoyed the symposium and their stay in Salvador.

October 2005
Aline Maria Santos Andrade
Flávio Morais de Assis Silva

Preface

Welcome to the proceedings of LADC 2005.

We are very proud of the high-quality program that LADC offered this year in Salvador.

It was our privilege to have the opportunity to select from such high-level papers as were submitted to LADC 2005. The profile of those submissions clearly shows that the previous (and first) LADC edition in 2003 was successful in setting a high-quality standard that all the prospective authors understood.

The 16 accepted papers, chosen from 39 submissions, laid out the guarantee of a technically very rewarding conference. The review process was very careful and selective, and we felt that our mission was that of strengthening the high-quality and international character of LADC. The accepted papers are from seven different countries, the majority of them from outside Latin America. The submitted papers had a similar profile, showing that LADC is clearly a conference that, in spite of the focus on Latin America, has a strong international visibility.

All papers were reviewed by four members of the Program Committee, and if needed by outside reviewers. The reviews were in general quite detailed, offering significant advice to the authors of accepted papers in preparing the final version, and to the authors of rejected papers in perfecting their work for a future submission. The acceptance decision was taken after a careful evaluation of all reviews, paying a very special attention to the reviews' content and not just the proposed numerical scores.

Finally, we would like to acknowledge the enthusiastic support of the LADC Steering Committee in all steps of this process, as well as of the Brazilian Computer Society for hosting the web tool supporting the paper submission and selection process. A word is also in order to our institutions, the Pontifícia Universidade Católica do Paraná and the Universidade de Coimbra, without whose support we would not have been able to perform this task. A special thanks goes also to William Sanders and Christof Fetzer for accepting to hold the keynote presentations.

Above all, we are confident that LADC 2005 will be remembered as a technically very rewarding conference, for the quality of both the papers and the discussions and contacts in Salvador.

October 2005

Carlos Maziero
João Gabriel Silva

Organizing Committee

General Co-chairs: Aline Maria Santos Andrade (UFBA, Brazil)
                             Flávio Morais de Assis Silva (UFBA, Brazil)

Program Co-chairs: Carlos Alberto Maziero (PUCPR, Brazil)
                             João Gabriel Silva (UCoimbra, Portugal)

Publication Chair: George Lima (UFBA, Brazil)

Publicity Chair: Luciano Porto Barreto (UFBA, Brazil)

Finance Chair: Sergio Gorender (UFBA, Brazil)

Local Arrangement Co-chairs: Marcela Santana (UFBA, Brazil)
                             Sandro Santos Andrade (UFBA, Brazil)

Registration Co-chairs: Frederico Barboza (UFBA, Brazil)
                             Ivo de Carvalho Peixinho (UFBA, Brazil)

Tutorial Co-chairs: Elias Procópio Duarte Jr. (UFPR, Brazil)
                             Sergio Rajsbaum (UNAM, Mexico)

Workshop Chair: Raul Ceretta Nunes (UFSM, Brazil)

WDAS Co-chairs: Herman Augusto Lepikson (UFBA, Brazil)
                             Leandro Buss Becker (UFSC, Brazil)

WTD Co-chairs: Avelino Zorzo (PUCRS, Brazil)
                             Ingrid Jansch-Pôrto (UFRGS, Brazil)
                             Fabíola Gonçalves P. Greve (UFBA, Brazil)

Steering Committee

Francisco Brasileiro, Brazil   Carlos Maziero, Brazil
Joni da Silva Fraga, Brazil   Sergio Rajsbaum, Mexico
Rogério de Lemos, UK   Taisy Silva Weber, Brazil
Raimundo Macêdo, Brazil   Flávio M. Assis Silva, Brazil
Eliane Martins (Chair), Brazil

LADC Program Committee

Pedro Mejia Alvarez, Mexico
Lorenzo Alvisi, USA
Pedro d'Argenio, Argentina
Jean Arlat, France
Marinho Barcellos, Brazil
Andrea Bondavalli, Italy
Francisco Brasileiro, Brazil
João B. Camargo Jr., Brazil
Ricardo Cayssials, Argentina
Jose Contreras, Chile
Mariela Curiel, Venezuela
Xavier Défago, Japan
Elmootazbellah Elnozahy, USA
Joni da Silva Fraga, Brazil
Paulo Lício de Geus, Brazil
Michel Hurfin, France
Ravi Iyer, USA
Ingrid Jansch-Pôrto, Brazil
Ricardo Jiménez-Peris, Spain

Jörg Kaiser, Germany
Johan Karlsson, Sweden
Kane Kim, USA
Jean-Claude Laprie, France
Rogério de Lemos, UK
Raimundo Macêdo, Brazil
José C. Maldonado, Brazil
Eliane Martins, Brazil
Fernando Pedone, Switzerland
Ravi Prakash, USA
Michel Raynal, France
Cecília M. Rubira, Brazil
William H. Sanders, USA
Richard Schlichting, USA
Paulo Veríssimo, Portugal
Pedro Gil Vicente, Spain
Raul Weber, Brazil
Taisy Weber, Brazil

LADC External Referees

Jorge Rady Almeida Jr., Brazil
Alysson Bessani, Brazil
José Eduardo Brandão, Brazil
Andrey Brito, Brazil
Lásaro Camargos, Brazil
Julien Cartigny, Japan
Mauro Fonseca, Brazil
Shashidhar Gandham, USA
Diogo Kropiwiec, Brazil
Srikant Kuppa, USA
Dorival Leão, Brazil
Lau Lung, Brazil
Paulo Marques, Portugal
Paulo Masiero, Brazil

Carlos Maziero, Brazil
Neeraj Mittal, USA
Mansoor Mohsin, USA
Felipe Pereira, Brazil
Lívia Sampaio, Brazil
Giuliana Santos, Brazil
Rodrigo Schmidt, Switzerland
Flávio Assis Silva, Brazil
João Gabriel Silva, Portugal
Henrique Silva, Portugal
Adenilso Simão, Brazil
Matthias Wiesmann, Japan

Organizer

Distributed Systems Laboratory (LaSiD),
Department of Computer Science (DCC),
Federal University of Bahia (UFBA)

Sponsor

Brazilian Computer Society (SBC)

In Co-operation with

IEEE TC on Fault-Tolerant Computing
IFIP Working Group 10.4 "Dependable Computing and Fault-Tolerance"
SADIO, Argentine Society for Informatics and Operations Research
SCCC, Chilean Computer Science Society
SMCC, Mexican Society for Computer Science

Table of Contents

Workshops

Tutorials

Probabilistic Validation of Computer System Survivability*

William H. Sanders

Donald Biggar Willett Professor of Engineering,
Dept. of Electrical and Computer Engineering,
Coordinated Science Laboratory and Information Trust Institute,
University of Illinois at Urbana-Champaign, USA
whs@uiuc.edu

There is a growing need for systems whose survivability in a specified use and/or attack environment can be assured with confidence. Many techniques have been proposed to validate individual components (e.g., formal methods) or a system as a whole (e.g., red teaming). However, no single technique can provide the breadth of evidence needed to validate a system with respect to high-level survivability requirements. To accomplish this, we propose an integrated validation procedure (IVP) that begins with the formulation of a specific survivability requirement R and determines whether a system is valid with respect to R. The IVP employs a top-down approach that methodically breaks the task of validation into manageable tasks, and for each task, applies techniques best suited to its accomplishment. These efforts can be largely independent, and the results, which complement and supplement each other, are integrated to provide a convincing assurance argument. We then illustrate the IVP by applying it to an intrusion-tolerant information system being developed by the U.S. Department of Defense. In addition to validating the system against high-level survivability requirements, we demonstrate the use of model-based validation techniques, as a part of the overall validation procedure, to guide the system's design by exploring different configurations and evaluating tradeoffs.

---

* This is joint work with Sankalp Singh, Adnan Agbaria, Fabrice Stevens, Tod Courtney, John F. Meyer, Partha Pal, and the rest of the DPASA project team. The author is grateful for this collaboration.

Timed Asynchronous Distributed Systems

Christof Fetzer

Technische Universität Dresden, Fakultät Informatik, Dresden, Germany
christof.fetzer@inf.tu-dresden.de
http://wwwse.inf.tu-dresden.de

The development of dependable distributed systems needs to be based on a proper foundation. This foundation is generally given in form of a system and failure model. The system model defines the semantics of basic services like process and message services of a distributed system. More advanced system services will be based on these basic services. The failure model specifies the likely failures of the basic services, i.e., these are the failures that the advanced system services need to cope with.

The objective of the system and failure model is the following. As long as the assumptions of the failure and system model are valid, a system has to guarantee its specification. However, if these assumptions are violated during run-time, the system specification might be violated. The probability that a dependable system violates its specification must be negligible. Therefore, the probability of the occurrence of failures which are not specified by the failure model must also be negligible.

In general, making stronger failure and system model assumptions simplifies the development of advanced system services. However, stronger assumptions increase the probability that the system and failure model assumptions are violated during run-time. Weaker assumptions reduce the probability of run-time violations but too weak assumptions will not permit a correct implementation of the system specification. Therefore, one needs to find assumptions that permit to implement the system specification while having a negligible probability of run-time violations.

The timed asynchronous system model [1] defines three basic services: a process, a clock and a communication service. All services are associated with appropriate failure assumptions. The process service has crash / performance semantics. The communication service has omission / performance semantics. The clock service is assumed to be failure free, i.e., each non-crashed process has access to a correct clock service.

The failure assumptions must be enforced. For example, clocks might fail. If the probability of a clock failure is not negligible in a certain system, one needs to enforce that clock failures are masked. For example, a clock failure could be transformed into a process crash failure to prevent that clock failures become visible. I will demonstrate how one can enforce the failure assumptions of the timed asynchronous system model using different model enforcement techniques. Using appropriate enforcement techniques the timed asynchronous system model is suitable even for dependable system with very stringent dependability requirements.

C.A. Maziero et al. (Eds.): LADC 2005, LNCS 3747, pp. 2–3, 2005.

Recently, we introduced a time-free model [2] that is based on the assumption that the average response times of non-crashed processes are finite to permit a deterministic solution of the consensus problem. I will show how this finite average response time can be added as an option to the timed asynchronous system model. I will discuss how this assumption can complement the traditional assumptions of the timed asynchronous system model, for example, in the domain of grid computing.

References

1. Flaviu Cristian and Christof Fetzer. The timed asynchronous distributed system model. *IEEE Transactions on Parallel and Distributed Systems*, pages 642–657, June 1999.
2. Christof Fetzer, Ulrich Schmid, and Martin Suesskraut. On the possibility of consensus in asynchronous systems with finite average response times. In *Proceedings of the 25th International Conference on Distributed Computing Systems (ICDCS 2005)*, 2005.

WLAN in Automation - More Than an Academic Exercise?

Edgar Nett

Otto-von-Guericke-Universitt Magdeburg,
Institut fr Verteilte Systeme,
Universittsplatz 2, 39106 Magdeburg, Germany
nett@ivs.cs.uni-magdeburg.de

Nowadays information technology (IT) is increasingly determining growth in the world of automation. After it changed hierarchies, structures and flows in the entire office world, it now covers all the sectors from the process and manufacturing industries to logistics and building automation. The communications capability of devices and continuous, transparent information routes are indispensable components of future-oriented automation concepts.

Today, two aspects determine new automation concepts. On the one hand they consist of distributed and component-oriented control structures, on the other hand the industrial automation and the IT of high-level management fields are growing more and more together (vertical integration). The expected benefits are

- Open communication from company management level to the field level (vertical and horizontal integration)
- Safe of investments through seamless integration of existing fieldbus systems
- Higher functional scope and performance as conventional fieldbus systems
- Simple and vendor independent plant wide engineering
- Extended range of applications: Remote Access, wireless communication

PROFINET possibly is the most advanced concept, especially for distributed automation standards. It is based on Ethernet and integrates existing fieldbus systems (in particular PROFIBUS) simply and without change. This is an important aspect for meeting the demand for consistency from the corporate management level to the field level. Furthermore, it represents a key contribution to providing the user with security for his investment in that existing parts of a system can be incorporated without needing to be changed.

Regarding real-time aspects, the Ethernet-based communication can be scaled along three levels:

1. TCP, UDP and IP for non-time-critical data, such as parameter assignment and configuration,
2. Real-Time (RT) for time-critical process data used in the field of factory automation and
3. Isochronous Real-Time (IRT) for particularly sophisticated demands, as for Motion Control applications.

C.A. Maziero et al. (Eds.): LADC 2005, LNCS 3747, pp. 4–8, 2005.

These rt levels conform nicely to with three different communication levels in automation systems realized by different bus systems.

At sensor/actuator level the signals of the binary sensors and actuators are transmitted via a sensor/actuator bus. Here, a particularly simple, low-cost installation technique, through which data and a 24-volt power supply for the end devices are transmitted using a common medium, is an important requirement. The data are transmitted purely cyclically.

At field level the distributed peripherals, such as I/O modules, measuring transducers, drive units, valves and operator terminals communicate with the automation systems via an efficient, real-time communication system. The transmission of the process data is effected cyclically, while alarms, parameters and diagnostic data also have to be transmitted acyclically if necessary. Fieldbuses like PROFIBUS meet these requirements and offer a transparent solution for manufacturing as well as for process automation.

At cell level, the programmable controllers such as PLC and IPC communicate with each other. The information flow requires large data packets and a large number of powerful communication functions. Smooth integration into company-wide communication systems, such as Intranet and Internet via TCP/IP and Ethernet are important requirements.

What is still missing in the convergence of IT and automation technologies is the integration of wireless communication. Even though this rises some tough challenges there are new application fields that drive a strong trend to deploy WLANs in industrial applications. Among these new applications fields mobile transport systems of all kinds are the most important and widely recognized sector. This sector spans from rail-guided baggage carriers that improve throughput and flexibility in airport baggage logistics, over warehouse systems with integrated transport entities, to AGVs and overhead monorail carriers that transport work pieces in assembly systems. In all these applications, providing wireless connectivity to the mobile entities promises a more detailed and up-to-date supervision and diagnosis, a more flexible control and an improved scalability.

The envisaged applications can be characterized w.r.t to the kind and tightness of control they exert via the wireless medium. The spectrum ranges from monitoring and diagnosis only (no control) over commissioning (task assignment), to autonomous or even centralized motion control. Depending on the kind of application, the WLAN will be used for the cell / production control layer with TCP / IP communication and no or soft real-time requirements, or the field bus layer with real-time requirements in the range from 100ms down to 10ms.

Using wireless communications in such demanding applications and environments poses some tough challenges. Besides fulfilling typical hardware requirements of industrial equipment, like DC voltage supply, standard industrial plugs, rugged housing with a sufficient protection against dust, water and heat, this mostly applies to the non-functional properties of the communication:

- Real-time. The envisaged applications are all subject to real-time requirements. How tight these requirements are depends on the kind of control tasks performed. Typically, a cyclic communication with a deterministic timing is required. Since this constitutes an end-to-end timing requirement between a mobile entity and another mobile entity or the cell / process controller, the roaming delays of the mobile entities have to be considered as messages may be lost or delayed while the source/destination station is in transition between two APs. Thus, fulfilling the real-time requirements not only means predictable and short medium access delays, but also a predictable and fast roaming.

- Reliability. Wireless media are by their very nature more error prone than wired ones. They are unshielded against EMI and suffer from different kinds of fading. Thus, measures have to be taken to achieve a reliability that meets the requirements of the applications, which in many cases are not designed to tolerate message losses. However, reliability measures such as retransmissions impact message delays and hence may conflict with the real-time requirements. The resulting trade-off has to be addressed when designing the reliability measures and when provisioning the networks for the application at hand.

- Availability. As WLANs become part of the control system, availability becomes crucial since unavailability of the wireless network may stop production and hence incur significant costs. Several kinds of measures must be considered to improve availability: fault avoidance measures to increase MTTF, like uninterruptible or redundant power supplies, fast diagnosis and easy (automatic at best) configuration and provisioning of replacement components to reduce MTTR; or measures that support active redundant WLAN deployments with fast fail over.

- Security. Again, the more the production relies on the underlying network, the more security becomes a key requirement. While physical security was considered a sufficient solution in many classical field bus and office systems, this is clearly not viable for wireless media because the physical access to the medium is hard to constrain. Therefore, measures have to be taken that achieve at least a level of security comparable to what the traditional systems do offer. While it is possible to employ sufficiently strong protection measures to achieve this goal, a key factor in industrial applications is also to consider the timing overhead these measures incur. Similar to reliability, the key challenge here is to achieve a sufficient security and at the same time fulfil the real-time requirements of the application.

Generally speaking, all the requirements can be captured under the notion of transparency. What people in automation would like to have is a communication channel that allows connecting mobile devices, but on a certain level looks like a traditional field bus system. This corresponds to the PROFINET idea to exchange the physical basis but maintain the properties. It should be noted, however, that this kind of transparency w.r.t the non-functional requirements means that the wireless cable replacement must fulfil the same timing and reliability requirements as the wired field bus, a challenging and ambitious objective.

During the last years a notable progress has been made towards the achievement of these goals: The IEEE has been and is still working on amendments to improve the non-functional properties of WLANs. IEEE 802.11i provides the required security features and actually steps beyond what is currently provided by wired links. However, the mechanisms employed incur a significant overhead and hence are not well suited for real-time applications. IEEE 802.11e will be adopted soon as an amendment to provide QoS on wireless media. While it specifies the air interface for resource reservation and allocation it intentionally leaves open the actual resource scheduling. Other task groups in IEEE 802.11 are working on reducing the roaming delays (802.11r) and improved monitoring and diagnosis interfaces (IEEE 802.11k). Furthermore, the IETF is in the process of standardizing a protocol for so-called Switched-WLAN architectures where part of the functionalities of the APs is delegated to a central controller, which is also in charge of configuration, provisioning and monitoring of the APs.

While the fast evolvement of new standards will provide the means (e.g authentication and encryption protocols, real-times MACs, measurement interfaces, protocols for centralized AP management, etc.) to address the requirements stated above it also

- brings about significant confusion regarding future development and whether investments spent today are secure. Actually, many of the standards (e.g. CAPWAP) are still under development and it is not clear which proposal will be adopted. Even for those standards which have already been implemented there are doubts, if there will be a sufficiently sustained support and availability of products.
- overburdens people who are not networking professionals and are now confronted with ever new technologies.

Besides a kind of settlement in the technological evolvement, what is needed with all the standardized technologies at hand is a coherently integrated solution that provides transparency of the underlying technology. Transparency means the solutions hides the intricacies of the underlying network from those engineers designing the plant and even more from the staff that will finally operate it. The talk will shortly explain what current standards and standardization efforts contribute to the achievement of this goal and what lines of development should be pursued to finally arrive at an integrated and transparent solutions. For example, such lines of development are:

- Scheduling of the wireless medium. This not only means scheduling the network access within a single cell. Rather, if infrastructure networks are considered, a global scheduling is required ensuring that roaming mobile entities will get their resource demands fulfilled with a sufficient probability. Such a scheduling should automatically be performed based on the information provided during the engineering of the plant control system and the resulting schedules should automatically be provisioned to the APs. Furthermore, the scheduling has to be integrated with other aspects such as capacity planning, transmission power control, and rate selection.

- An application-dependent selection of reliability measures (e.g. FEC, ARQ, no ack) and parameters (e.g. retry limits) should be supported in a way that does not require application engineers to be networking experts. Furthermore, making this choice has to be integrated with the scheduling and capacity planning.
- Diagnosis components that not only provide large amounts of detailed data but support operators in comparing those data against real-time and reliability requirements and underlying load and capacity assumptions.
- A centralized management of users / devices and their credentials that incurs a minimum of additional maintenance effort and roaming overhead and is applicable for mobile devices with limited processing resources. " Optimization of roaming delays based on centrally available information like site surveys, client positions, load information etc.

Acknowledgement

I would like to thank very much Dr. Stefan Schemmer from rt.solutions.de for his valuable contributions from a business perspective.

Using Stratified Sampling for Fault Injection

Regina Lúcia O. de Moraes[1], Eliane Martins[2], Elaine C. Catapani Poletti[1],
and Naaliel Vicente Mendes[1]

[1] Superior Centre of Technological Education (CESET),
State University of Campinas (UNICAMP)
{regina, elainec, naalielb}@ceset.unicamp.br
Phone: +55 19 3788-5872/ Fax: +55 19 3404-7164
[2] Institute of Computing (IC),
State University of Campinas (UNICAMP)
eliane@ic.unicamp.br
Phone: +55 19 3404-7165/ Fax: +55 19 3788-5847

Abstract. In a previous work we validated an ODBMS component injecting errors in the application's interface. The aim was to observe the robustness of the component when the application that interacted with it failed. In this work we tackle the injection of errors directly into the interfaces among the target component's classes. As the component under test has several classes, we use stratified sampling to reduce the amount of injections without losing the ability to detect faults. Strata are defined based on a complexity metric, Weighted Methods in a Class – WMC. Experiments show that this metric alone is not sufficient to select strata for testing purposes.

1 Introduction

Increased pressures on time and money make component-based software development a current trend in constructing new systems. The development of a system that is an integration of several Off-The-Shelf (OTS) components brings hypothetical benefits, such as system quality enhancement, since the components are used in other systems, and time and money savings, since the source code does not need to be rewritten. Moreover, components and component-based system validation is still a challenge.

The difficulty stems from the degree of knowledge that developers and users have about the component [2] [18]. When developing a component, the developer cannot picture every possible use this component may have in the future. Component users do not know the acquired component's quality level, and even if it is known, there is no guarantee that the component will present the same quality level when used in a new context. Furthermore, the use of high-quality components does not guarantee that the overall system will have high quality, due to the complexity of interaction among components [19].

Component validation is therefore a very important task. It allows us to determine whether the component provides the expected services, and to check whether it does not present unexpected harmful behaviour. Fault injection is a useful technique in which faults or errors are deliberately introduced into a system in order to observe its

C.A. Maziero et al. (Eds.): LADC 2005, LNCS 3747, pp. 9–19, 2005.
© Springer-Verlag Berlin Heidelberg 2005

behaviour and thus better understand how robust the software is, how efficient is it when recovering its normal execution after a non-successful transaction, and the impact of its detection and recovery mechanisms on the application's performance.

In a previous work [10] we validated the Object-Oriented Database Management System (ODBMS) Ozone [12], an OTS component, aimed at evaluating its robustness in the presence of errors originated in the application. The benchmark Wisconsin OO7 was our target application. The Jaca tool [9] was used to inject errors at the interface between OO7 and Ozone. A risk-based strategy [1] [15] was proposed and applied for the selection of OO7 classes in which to inject. In that work, we injected in the selected classes and in all OO7 classes to evaluate the effectiveness of the strategy and to compare the results.

In this work the component under test has several classes and injection in all classes would be too time-consuming. We consider stratified sample and ratio theory to determine the number of elements that allows us to get a confidence in the sample. The number of elements taken from each stratum conserves the same proportion presented by the set of all component classes. One difficulty with stratified sampling is the determination of the strata. In order to address this difficulty, in this work we present a risk-based strategy used in [10]. Section 4 briefly presents this strategy. The reminder of the paper is organized as follows. Ozone as well as OO7 are presented in Section 2. Fault Injection fundamentals as well as some related works are shortly presented in Section 3. The results of the stratified sampling strategy applied to the case study are presented in Section 5. Finally, Section 6 concludes this work.

2 Case Study Description

The case study used for strategy testing is a system composed by two main components, an ODBMS called Ozone and the OO7, a benchmark used to evaluate ODBMS performance. In this experiment the benchmark is seen as the application responsible for the activation of injected faults and the propagation of errors to the component under test, the Ozone database.

2.1 The Target Component

Our target component is an object database management system (ODBMS) called Ozone [12]. Written in Java, it allows Java objects in a transactional environment to persist according to the structure defined by the application. Based on client-server architecture, clients connect to the database using sockets with a RMI protocol. To guarantee a unique instance in the server, Ozone uses "proxy" objects that can be seen as a persistent reference. These proxies are generated by the Ozone Post Processor (OPP) as a result of two linked files, the class file and an external interface that is created for each class. The experiments performed in this work use a local configuration, but Ozone can also be used in a distributed architecture.

2.2 The Target Application

We use Wisconsin OO7 [3] as a benchmark application to activate the target component. Wisconsin OO7 was found in Ozone's website [12] and was originally used to evaluate the ODBMS performance.

The main component of the benchmark is a set of *composite parts*. Each *composite part* has a *document* object and a graph of associated *atomic parts*. A set of all *composite parts* forms the *design library*. *Assembly* objects are more complex structures, which may be composed of *composite parts* (base assembly) or other *assembly* objects (complex assembly). These assemblies are organized in hierarchies; each of which constitutes a *module*. There is a *manual* to document each *module*.

The Ozone's version of OO7 implements three main functionalities, one to store objects and create an assembly hierarchy (create), one to search root objects (query match), and another to traverse the composite part objects' hierarchy (query traversal) [3]. To check the ACID properties (Atomicity, Consistency, Isolation and Durability) we implement extra functionalities that are based on TPC-C benchmark [16] and OO7 specification [3]. The extra functionalities are used to check the database state before and after an experiment execution. They are described in more detail in [10]. In short, to verify atomicity we use OO7 queries and other queries created, and then compare the stored data before and after fault injection. To check consistency, a query is performed to verify whether the new data stored in the database is in accordance with OO7 specification. Durability is checked by disconnecting and connecting the database and comparing its state through the queries results. As the experiments are performed in a local machine, isolation is not checked.

Among three possible sizes of the database created by OO7, this work uses the smallest one, which contains one assembly hierarchy with seven levels, composed of two other assemblies. Composite parts with a total of 500 per module define the assemblies in the lowest level. Each composite part contains 20 atomic parts, comprising a total of 10,000 atomic parts [3].

3 Software Fault Injection

Fault injection is a technique that simulates anomalies by introducing faults into the systems under test and then observing their behaviour. Among the various existing fault injection approaches (see [7] for an overview), software fault injection has been widely adopted. It can be used to simulate internal faults, as well as faults that occur in external components interacting through interfaces [18]. One approach of software-implemented fault injection consists of injecting anomalous input data that comes into the software through its interface [20]. This study uses this approach, allowing software acquirers to determine its robustness. The software can be stated as robust if it is fed by anomalous input without propagating the error that may cause a failure. This demonstrates that the software can produce dependable service even in presence of an aggressive external environment [20].

To apply this approach, a tool is needed to inject faults during runtime. We use Jaca [9], a software-implemented fault injection tool that offers mechanisms for the injection of interface faults in Java object-oriented systems. Jaca is source code independent, allowing the validation of a system that may be composed of multiple third-parties components. Jaca's current version can affect the public interface of a component by altering values of attributes, method parameters and returns. These values must be simple (integer, float and Boolean), strings or objects. Jaca is described in more detail in [9].

A similar approach is presented by Ballista [8] and Mafalda tools [6], but in those cases the errors are injected in the parameters of operating system calls instead of the components interfaces. As with the Mafalda tool, we also consider the errors published by Ballista approach. TAMMER [5] is another similar work in which the injection of interface faults is used to observe fault propagation focused on code coverage, while in our work we are interested in the exceptions raised as well as whether these exceptions cause the whole system to fail. Unlike TAMMER, we do not need the source code.

4   Characterization of the Experiments

The target component Ozone contains 430 classes. In this case, injecting all classes would be a hard and unpractical work. We need a way to select a sample of classes to be injected. For this purpose we use stratified sampling. Stratified sampling and partition testing are presented in [13] to estimate reliability. In our work we use them to characterize the strata in order to test the robustness of a component-based system.

To use stratified sampling, the steps needed to define an experiment are the following: (1) define the stratification criteria and categorize Ozone classes in each stratum; (2) calculate the sample size; (3) apply the theory of proportion to determine the sample size for each stratum; (4) select the classes that belong to each stratum in the sample; (5) characterize the fault injection campaign.

4.1   Definition of the Stratification Criteria and the Strata

Stratified sampling is a sample technique that divides, based on any criteria, the population into strata and then associates another method to select the elements that should compose the sample. Partition testing can be considered a kind of stratified sampling, in which a system input domain is divided into partitions according to operational behaviour, and one input from each stratum is selected. One difficulty in stratified sampling is defining the stratum. In this work we use a risk-based strategy to determine the strata. In a previous work we used a set of complexity metrics, namely, the CK metrics suite, to determine class complexity [4]. According to a pre-specified threshold for each metric obtained in an experimental study with several real world classes [15], we were able to define the classes with high complexity, i.e., those for which one or more metric values lie outside the threshold. In this study we select one metric of the CK suite, the Weighted Method for a Class, or WMC. The WMC metric represents the complexity of a class in terms of the number of its methods and their complexities, and thus it is reasonable to consider them as a first choice for our assessment. The assumption is that the higher the WMC, the higher the error proneness of the class, making it a good candidate for fault injection. Thus, we calculated the WMC metric of all Ozone classes.

Based on the WMC metric obtained, we classify all the Ozone's classes and separate them in two strata according to the WMC metric thresholds: (S1) WMC metric is equal or smaller; (S2) WMC metric is greater [15].

4.2 Calculating the Sample Size

To estimate the sample size we need to determine the percentage of success and non-success, the confidence level and the error tolerance.

$$n = \frac{\left(Z_{\alpha/2}\right)^2 \cdot \hat{p} \cdot \hat{q}}{E^2} = 30{,}1181 \equiv 31 \text{ classes}$$

Where:

$E \equiv 0.05$ (5% - error tolerance)

$Z_{\alpha/2} \equiv 1.96$ (critical value related to the reliability on the failure ratio - 95%)

$\hat{p} \equiv 0.02$ (failure ratio based on the previous experiments = 45 failures / 2700 experiments)

$\hat{q} = 1 - \hat{p} = 0.98$

Fig. 1. Sample Size Estimation [17]

From the failure ratio of previous experiments [10], 45 out of 2700 experiments performed resulted in failure. In this way, the value of $\hat{p}$ is 45/2700, which is approximately 2%; thus, the complementary value, $\hat{q}$ is 98%. The confidence level considered is 95%, which implies a critical value of 1.96 and an error tolerance of 5% (complementary percentage related to 95%). Based on these values we obtained a sample size of 31 classes [17]. Figure 1 presents the sample size estimator.

4.3 Obtaining the Sample Size for Each Stratum

Given that N_{S1} represents the number of classes in stratum S1, N_{S2} the number of classes in S2, N the number of Ozone classes and n the estimated sample size, then the sample size (n_{Sx}) for a stratum (x) is given by: $n_{Sx} \equiv N_{Sx} / N * n$ and $\Sigma\ n_{Sx} = n$, according to the theory of proportions mentioned in Section 4. Since stratum S1 represents 89% of Ozone classes and stratum S2 11% of the total of classes, and considering a sample size of 31 classes, we need to select 27 classes from stratum S1 and 4 classes from stratum S2.

4.4 Selecting the Classes in Each Stratum

To select the classes to be sampled, we rank the classes in each stratum (Sx) in a decreasing order based on the WMC metric. Then we take n_{Sx} classes from the top WMC. We also take into account the class's visibility since we can inject only in a public class. Among the top classes of stratum S1 and stratum S2 there are classes in which is not possible to inject due to technical restrictions (all methods are protected).

So we consider the next one in the rank. From now on, we inject firstly into the classes that belong to the stratum S2, secondly into all classes of the sample, and only then compare the results. To confirm our strategy, no major different failures should occur when we compare both results.

4.5 Characterization of the Fault Injection Campaign

A fault injection campaign is characterized by a faultload, a workload and readouts to be collected. A faultload describes the set of faults that are going to be inserted in the target system, defined according to the fault representativeness and the established fault selection criteria [18]. A faultload is determined by a fault location, a fault's type, triggering condition, repetition pattern and injection start. These elements can be described as follows:

(i)   *Fault Location*: In this work we inject interface faults [21].
(ii)   *Fault's Type:* Corruption of the parameters and returned values, replacing them with invalid values, combining the Ballista approach [8] with boundary value testing [14] (based on the system's specification).
(iii)   *Triggering Condition*: interception of operation calls at the component interfaces.
(iv)   *Repetition Pattern*: the frequency of the injection (permanent, intermittent or transient).
(v)   *Injection Start*: how many times an operation must be called before the first injection.

The values to be injected are based on Ballista approach for robustness tests, together with the ones proposed in [18]. Table 1 presents these values, which should be chosen according to the parameters' or returned values' data type.

Table 1. Values to Inject based on the Ballista Approach

Data Type	Values to Inject
Integer	0, 1, -1, MinInt, MaxInt, neighbour value (current value $\pm$ 1)
Real Floating Point	0, 1, -1, DBLMin, DBLMax, neighbour value (current value * 0.95 or * 1.05)
Boolean	inversion of state (true -> false; false ->true)
String	Null

The workload is the based program(s) that run(s) on the system when the experiment is conducted [18]. In this study the workload is the Benchmark Wisconsin OO7.

The readouts are collected from several sources: (i) From Ozone's interface, we extract the number of stored clusters, and the exceptions thrown by Ozone that were not treated. (ii) From Benchmark's interface on Jaca, we extract exceptions thrown that are not treated by the application. (iii) From Ozone's log, we take out data that are not similar to those of Ozone's interface. (iv) From Jaca's log, we extract specification of the injected faults and exceptions raised. (v) From stored data, we determine whether database consistency is guaranteed and we perform the existent queries to verify if all committed transactions were stored in the database. If an interruption occurs, we need to verify the stored data to certify the non-residual data. These outcomes are used in this study to characterize Ozone's behaviour in the presence of faults.

Ozone's behaviour can be characterized as follows: I) is the ideal case, in which both OO7 and Ozone have normal termination and the database created is in a consistent state. EXC OO7) is an exception generated at OO7 as a consequence of fault injection, but Ozone has normal termination and the database created is in a consistent state. This case characterizes the robustness of Ozone with respect to application failures. EXC OZ) is an exception thrown by Ozone, which terminates abnormally but the database created is in a consistent state. N) is used when Ozone terminates normally but the database created is in an inconsistent state, which violates the ACID properties. Finally, A) occurs when Ozone terminates abnormally and the database created is in an inconsistent state. Types N and A characterize failure of the database manager, in that it allows stored data to be corrupted as a result of non-successful transaction.

An error is said to have been tolerated when the system does not crash and the ACID properties are kept; a failure occurs when the system crashes or the ACID properties are not kept. A non-effective error is an error that causes no change in the system, and an error is considered non-detected when the system does not perceive the occurrence of an error and a failure.

5   Experimental Results

5.1 Strata Definition

Table 2 presents the selected Ozone classes in which to inject according to the strategy described in Section 4. As described in Section 4.3, the sample size should be composed by 27 classes from stratum S1 and 4 classes from stratum S2. The classes JavaCodeAnalyzer, Table and NumberLineEmitter are not considered since it is impossible to inject into them due to technical restrictions (all its methods are protected), as explained in Section 4.4.

The values below the metric's name (between parentheses) indicate the ideal and the maximum acceptable values of each metrics for Java applications (threshold values) [15]. The signalled classes are those with WMC metric greater than the maximum acceptable values.

Table 2. Selection of Ozone's classes to be injected

Class	WMC (25;40)[1]	Stratum	Can Inject into this Class?
SAXChunkProducer	133	S2	YES
WizardStore	117	S2	YES
JavaCodeAnalyzer	113	S2	NO
ProxyGenerator	111	S2	YES
NodeImpl	108	S2	YES
HTMLTableRowElementImpl	39	S1	YES
ParamEntity	38	S1	YES
OzoneODMGTransaction	38	S1	YES
CDHelper	38	S1	YES
CollectionImpl	37	S1	YES
CXMLContentHandler	37	S1	YES
HTMLObjectElementImpl	36	S1	YES
DbCacheChunk	35	S1	YES
SimpleArrayList	35	S1	YES
Table	35	S1	NO
AbsoluteLayout	34	S1	YES
DatabaseImpl	34	S1	YES
OzoneXAResource	34	S1	YES
CharacterDataImpl	33	S1	YES
CollectionImpl	33	S1	YES
DxAbstractCollection	33	S1	YES
HTMLElementImpl	33	S1	YES
NumberLinesEmitter	33	S1	NO
OPP	3	S1	YES
BLOB	32	S1	YES
DxMultiMap	32	S1	YES
HashtableContentHandler	31	S1	YES
AbstractObjectContainer	30	S1	YES
AdminObjectContainer	30	S1	YES
DocumentImpl	29	S1	YES
HTMLAnchorElementImpl	29	S1	YES
HTMLSelectElementImpl	29	S1	YES
Enh Properties	28	S1	YES
ExternalTransaction	28	S1	YES

5.2 Fault Injection Campaign

A total of 31 injection points with integer, long, string and objects data type are injected. We also vary the repetition pattern and the start time. Table 3 resumes the campaign.

---

[1] Ideal and acceptable limit values according to [15]. These values were obtained in tests conducted over a period of three years, in which over 20,000 Java classes where collected and analyzed.

Table 3. The Campaign Experiments' Distribution

Classes of Experiments	Start Time	Repetition Pattern	Number of Parameters/ Return Values Injected in S1	Number of Parameters/ Return Values Injected in S2	Total Injection
O1P	First Occurrence	Permanent	31	4	35
O1T		Transient	31	4	35
O1I		Intermittent	31	4	35
O2P	After First Occurrence	Permanent	31	4	35
O2T		Transient	31	4	35
O2I		Intermittent	31	4	35
	Total N° of Experiments		186	24	210

5.3 Experimental Results Analysis

From a total of 210 injections, 24 are performed on stratum S2 and 186 on S1. On stratum S2, 20 injections are type I (which cause no violation on Ozone's behaviour nor on stored data), and 4 are type EXC OO7 (which are tolerated by the system, causing the execution as well as all the queries performed to terminate normally). On stratum S1, 180 injections are type I. Among the other injections one of them is type A (which did not terminate normally and impacted the system leading it to a failure); the other five injections are type N (which presented no abnormality in Ozone's

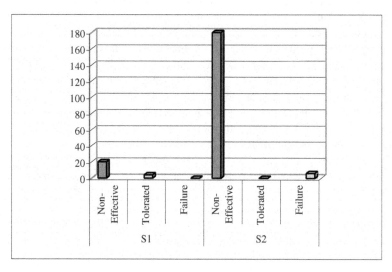

Fig. 2. Strata's Results

interface but the queries could not be performed, pointing that the stored data was corrupted, which in turn violated the ACID properties). To check stored data, we make a new connection with the database, invoke a query match (in which the root objects are checked) and a query traversal (which allows us to check the assembly hierarchy). Figure 2 presents the results for each stratum.

These results show that although the classes with higher WMC are more prone to errors than the ones with lower WMC, the impact of their faults in the system is not severe. This leads us to consider, in further experiments, other factors to define the strata.

6   Conclusions and Future Work

This work presents the use of stratified sampling for robustness testing purposes. The idea is to select components to inject in each stratum, instead of randomly selecting them from the whole set of system components. To define the strata, we use a complexity metric, WMC. The components are in fact divided into two strata: one for component with higher WMC value than the threshold value of this metric, and the other for lower WMC value than the same threshold value. We apply the approach for testing a database component, Ozone.

We perform experiments to evaluate the robustness of an *off-the-shelf* (OTS) component. Using a fault injection technique, we inject errors at chosen interfaces into Ozone.

The results show that Ozone's behaviour was different for each stratum, as expected. However, differently from our expectations, the stratum containing the classes with higher WMC does not produce the most severe failures; they do not cause database corruption.

In a previous work [10], the classes were selected according to several objected-oriented metrics selecting the classes with higher risk. In that work, the risk depended on various factors; among them the WMC metric.

This work is based on a single metric, the WMC, to select the Ozone classes in which to inject. The results show that the exclusive use of this metric is not sufficient to choose the strata. Other factors should be taken into account. For example, in the aforementioned work [11], we analyse the dependences among Ozone and OO7 classes. The results obtained are more promising results, highlighting that the dependence is more important in [10] than the WMC metric. Furthermore, the methods that implement a critical function in the system must be considered as a selection criterion, as shown in [10].

Further experiments are envisaged to define other criteria for stratification. As a long term goal, we intend to use stratified sampling to obtain inferences about a system's reliability.

Acknowledgment. This research is partly supported by CNPq – Brazil's National Council for Scientific and Technological Development – through the ACERTE project.

References

Bach, J.: Heuristic risk-based testing. Software Testing and Quality Engineering Magazine, (1999)

Beydeda, S., Volker, G.: State of the art in testing components. In: Proc. Of the International Conference on Quality Software, (2003)

Carey, M. J., DeWitt, D. J., Naughton, J. F.: The OO7 Benchmark. http://www.columbia.edu/, (1994), recovered February (2005)

Chidamber, K.: Principal Components of Orthogonal Object-Oriented Metrics. http://satc.gsfc.nasa.gov, (1994), recovered November (2004)

de Millo, R. A., Li, T., Mathur, A. P.: Architecture of TAMER: A Tool for dependability analysis of distributed fault-tolerant systems. Purdue University, (1994)

Fabre, J-C, Rodriguez, M., Arlat, J., Sizum, J-M.: Building dependable COTS microkernel-based systems using MAFALDA. In: Proc. of 2000 Pacific Rim International Symposium on Dependable Computing - PRDC'00, Los Angeles, USA, (2000)

Hsueh, M. C., Tsai, T., Iyer, R.: Fault Injection Techniques and Tools. In: *IEEE Computer*, (1997), pp. 75-82

Koopman, P., Siewiorek, D., DeVale, K., DeVale, J., Fernsler, K., Guttendorf, D., Kropp, N., Pan, J., Shelton, C., Shi, Y. Ballista Project : COTS Software Robustness Testing. Carnegie Mellon University, http://www.ece.cmu.edu/~koopman/ballista/ (2003)

Martins, E., Rubira, C. M. F., Leme N.G.M.: Jaca: A reflective fault injection tool based on patterns. In: Proc. of the 2002 Intern Conference on Dependable Systems & Networks, Washington D.C. USA, Vol. 23(267), (2002), pp. 483-487

Moraes, R., Martins, E.: A Strategy for Validating an ODBMS Component Using a High-Level Software Fault Injection Tool. In: Proc. of the First Latin-American Symposium, LADC 2003, pages 56-68, São Paulo, Brazil, (2003)

Moraes, R., Martins, E., Mendes, N.: Fault Injection Approach based on Dependence Analysis. In: Proc. of the First International Workshop on Testing and Quality Assurance for Component-Based Systems – TQACBS, (2005)

Ozone, Object Oriented Database Management System, www.ozone-db.org/, (2004)

Podgurski, A., Yang, C.: Partition Testing, Stratified Sampling and Cluster Analysis. In: Proc.of the 1st ACM SIGSOFT symposium on Foundations of software engineering. pp. 169-181, Los Angeles, USA, (1993)

Pressman, R. S.: Software Engineering a Practitioner Approach, 4th edition. Mc Graw Hill1, (1997)

Rosenberg, L., Stapko, R., Gallo, A.: Risk-based Object Oriented Testing. In: Proc. 13th International Software / Internet Quality Week (QW2000), San Francisco, California USA, (2000)

Transaction Processing Performance Council "TPC-C – Benchmarks". http://www.tpc.org/tpcc/default.asp, (2005)

Triola, M. F.: Introcução a Estatística, 7th Edition. LTC Editor, Rio de Janeiro, (1999) (in portuguese)

Voas, J., McGraw, G.: Software Fault Injection: Inoculating Programs against Errors. John Wiley & Sons, New York, EUA, (1998)

Voas, J. M., Charron, F., McGraw, G., Miller, K., Friedman, M.: Predicting how Badly Good Software can Behave.In: *IEEE Software*, (1997), pp. 73–83

Voas, J.: Marrying Software Fault Injection Technology Results with Software Reliability Growth Models. Fast Abstract ISSRE 2003, Chillarege Press, (2003)

Voas, J.: An Approach to Certifying Off-the-Shelf Software Components. In: IEEE Computer, 31(6), (1998), pp. 53-59

A Methodology for the Automated Identification of Buffer Overflow Vulnerabilities in Executable Software Without Source-Code

João Durães[1] and Henrique Madeira[2]

[1] ISEC/CISUC - Polytechnic Institute of Coimbra,
3030 Coimbra, Portugal
jduraes@dei.uc.pt
http://www.cisuc.uc.pt/view_member.php?id_m=80
[2] DEI/CISUC - University of Coimbra,
3030 Coimbra, Portugal
henrique@dei.uc.pt
http://www.cisuc.uc.pt/view_member.php?id_m=63

Abstract. This paper presents a methodology for the automated detection of buffer overflow vulnerabilities in executable software. Buffer overflow exploitation has been used by hackers to breach security or simply to crash computer systems. The mere presence inside the software code of a vulnerability that allows for buffer overflow exploitations presents a serious risk. So far, all methodologies devised to mitigate this problem assume source code availability or prior knowledge on vulnerable functions. Our methodology removes this dependency and allows the analysis of executable code without any knowledge about its internal structure. This independence is fundamental for relevant scenarios such as COTS selection during system integration (for which source code is usually not available), and the definition of attackloads for dependability benchmarking.

1 Introduction

Buffer overflow exploitation is currently a major cause of security breaches in software systems. From January 2004 to March 2005 at least 41.7% of the reported security holes were related to buffer overflow exploits [1] (possibly even more as some of the exploits were not detailed and may also be related to a buffer overflow).

This problem is not recent: the well known internet worm of 1988 was based on buffer overflow exploitation [2]. Despite the age of this problem, new cases of security breaches based on this exploitation keep appearing regularly (e.g., the recent Windows JPEG GDI+ case [3]). The omnipresence of this type of exploit is due to the fact that it is based on a relatively simple software weakness (i.e., one that is prone to exist) and much legacy code contains such weaknesses (e.g., some C library functions).

The consequences of a well-succeeded buffer overflow attack typically include the remote access to a root/administrator shell leading to all kinds of security breaches such as information theft and user impersonation. Less serious consequences typically

C.A. Maziero et al. (Eds.): LADC 2005, LNCS 3747, pp. 20–34, 2005.

imply system crash and system unavailability, leading to possible business losses. Thus, software containing weaknesses enabling this kind of attack represents a serious risk.

So far, research on this topic has not produced a methodology for the identification of buffer overflow weaknesses in executable software without the need of source code or any other previous knowledge on the software. Considering the current industry trend of COTS-based system development, software modules containing vulnerabilities leading to buffer overflow attacks are most likely included in deployed systems and in general system integrators are not aware of this weakness, as COTS source code is often not available. Thus, the detection of buffer overflow vulnerabilities directly in the executable code is particularly relevant.

The successful identification of executable software containing weaknesses leading to buffer overflow can be useful in a variety of situations:

- Decision making when choosing among different alternative COTS for system integration.
- Wrapper development for the vulnerable software.
- Definition of attackloads for dependability benchmarking.

This paper proposes a new methodology for automated discovery of software weaknesses that may lead to buffer overflow exploitation. The proposed methodology has the following advantages over previous approaches:

- It does not depend on source code availability.
- It does not require prior knowledge about weak library functions.

The paper structure is as follows: in the next section we describe prior work related to buffer overrun prevention and detection. Section 3 presents an overview of the most common and dangerous buffer overrun attacks. Section 4 presents our methodology. Section 5 discusses the methodology portability, and section 6 presents several applications scenarios. Section 7 concludes the paper.

2   Related Work

Previous research on this topic has provided several approaches to mitigate the problem of buffer overflow vulnerabilities. The RAD tool [4] uses compiler extensions that automatically insert protections in the source code. StackGuard [5] uses a similar approach. These approaches have strong limitations, as they can only be used within the development team and they are useless in a COTS-based software development scenario.

[6] uses source code static analysis aimed at the identification and correction of weak spots. This approach is also limited to the development team and offers no protection against potential buffer overflow vulnerabilities in executable software provided by third-parties.

In [7] a run-time protection using robust implementation of previously known weak libraries is proposed. This proposal offers some protection against third-party developed software provided that there is prior knowledge regarding known weaknesses. Unfortunately, this is not often the case, as the weaknesses are discovered after a system has been compromised (that is the scenario of the internet worms based on software vulnerabilities).

Another approach uses fault injection to insert attack code into the observed software to measure its vulnerability (e.g., FIST [8]). However this approach modifies the software being evaluated which turns the conclusions obtained afterwards quite dubious. The tool LibVerify referred in [7] can be used to detect vulnerabilities in executable software modules; however it also relies on the modification of the observed code.

In [9] it was shown that reverse engineering techniques can be used to discover hidden vulnerabilities without source code. However, no systematic methodology or tool was proposed.

TaintCheck [10] is a tool that analyses the propagation of unsafe data through the system in an attempt to discover attacks during runtime. Although this tool does not require source code availability, it is limited to runtime detection and cannot be used for attack avoidance purposes. It also causes performance overhead in the target software.

SAFE [11] is a tool that performs static analysis of executable software to detect the presence of malicious code based on instruction patterns and signatures. Although the techniques employed in the analysis of the executable code resemble reverse-engineering techniques and are independent of source code, they do not target the detection of buffer overrun weaknesses.

In [12] is presented another technique based on static analysis of executable software and reverse engineering techniques. Although independent from source code availability, it specifically tied to vulnerabilities surrounding unsafe use of the C function *sprintf*.

3   Buffer Overflow and Stack Smashing

The simplest and most common form of buffer overflow attack is known as stack smashing [13]. This attack consists of supplying more data to a given software module than the amount of data it can store in its internal stack-resident buffer. While storing the data, the software module writes beyond the end the buffer and eventually overwriting the return address, which is also stored in the stack (see Figure 1). The value that replaces the return address is usually the address of a small code portion that performs an attack action (usually the spawning of a shell). When the function terminates the return address is fetched from the stack and the execution jumps to the location specified by the fake address which points to malicious code crafted by the attacker.

The malicious code is usually supplied as part of the data being fed to the buffer. Thus, software modules that use input data are strong candidates for stack smashing attempts. If the attack is successful, the attacker code is executed in context of the running process. This means that the malicious code is executed with the identity and privileges of the process owner. If the process belongs to a system level service (e.g., a web server) its privileges will be typically high. Thus, depending on the target nature, the malicious code has potentially complete access to system.

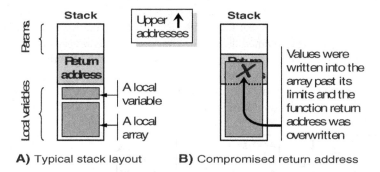

A) Typical stack layout   **B)** Compromised return address

Fig. 1. Stack smashing overview. The typical stack layout consists of the function local variables followed by the function return address (*left*). If one of the local variables is overwritten with more bytes than those it can hold, the return address may be corrupted and cause the execution flow to be hijacked to arbitrary code (*right*). Buffers (arrays) are a type of variable which are prone to be filled with more data than they can hold.

Stack smashing is based on two simple factors: a) the stack grows towards the lower addresses, and b) normal operations performed on variables, such as filling a buffer, are carried towards the upper addresses. Thus the single basic software weakness required to enable a stack smashing attack is the lack of explicit checks for the limits of buffers. Unfortunately, programmers often use assumptions regarding the maximum space needed for a given byte sequence (e.g., a string) and often omit checks to validate those assumptions.

All the platforms where the stack grows towards the lower addresses are susceptible to stack smashing. This type of architecture represents the vast majority of platforms used nowadays, which makes the problem of stack smashing omnipresent.

4 Methodology Proposal

Our methodology is based on the fact that the weakness necessary for buffer overrun exploitation can be detected through the automated analysis of the machine-code instructions of the observed software (in fact, exploits are sometimes discovered through reverse engineering, although in a non-automated fashion [9]).

The methodology is composed by two main phases. During the first phase we use the knowledge about the weakness leading to buffer overflow as a search pattern and analyze the executable code to identify probable locations containing this weakness. This analysis is performed in a similar way as a software fault emulation technique developed in our group [14, 15]. The main result of the first phase is the characterization of the software functions (modules) as "safe" or "suspect".

The second phase of our methodology consists of a robustness test of the "suspect" functions in order to confirm that the suspect functions are indeed vulnerable (it may be the case that some suspect functions are just "false positive"). This test consists on supplying values to the function parameters that are likely to cause a buffer overflow. This is similar to API robustness testing methods [16,17]. However, the values supplied to the parameters are specifically intended to cause buffer overruns and are based on the knowledge obtained during the first phase. This drastically reduces the time needed to run the experiments and increases the chances of activating existing weaknesses.

Figure 2 presents a visual overview of the methodology. We detail the two steps of our methodology in the following sub-sections.

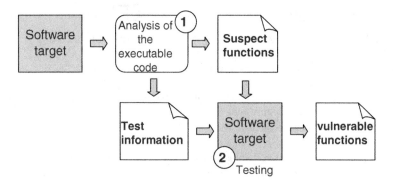

Fig. 2. Methodology overview. The first step consists of the analysis of the observed executable code to identify the existence of buffers in the stack and infer how such buffers are manipulated. This analysis provides a preliminary classification of clear/suspect functions and information regarding which values should be used to test the suspect function during the second step.

4.1   Phase 1 – Identification of Potential Weakness Locations

The first phase comprises the analysis of the low-level code of the software being observed. In this context low-level code means the executable file format (e.g., PE format in the Wintel platform). The methodology only requires the knowledge of the software starting address and does not need any meta-data information or compiler-generated debug information.

The primary objective of phase 1 is to identify and characterize the software modules regarding its usage of stack space and the relationship existing between each module. In this context, module means "function" or "procedure". To avoid conflict with other common usages of the term "module" such as in "system module" or even "COTS", we refer from now on the software functions and procedures as simply "functions."

During this step the low-level code is analyzed and the instruction sequence of its functions interpreted (within the limits of an automated analysis). The goal is to locate the code signatures of the programming constructs that are related to buffer use and buffer-limit check omission. Note that the process of "signature location" should not be interpreted as a simple byte-scan of the observed software. In fact, it is a much more complex process during which the meaning of the low-level instructions is taken into account to detect the distinctive marks of the programming constructs related to the weaknesses required for buffer overflow and stack smashing attacks.

The two distinctive constructs required for a successful smash stacking attack are:

- The existence of a buffer stored in the stack.
- Instructions to fill the buffer without properly checking the buffer boundary.

It is worth noting that the buffer and the instructions to fill it with data do not need to be located in the same function: the function containing a buffer may call a second function supplying the buffer to be filled within that second function. In fact, this is quite a common scenario (see example in Figure 3). Thus the relationship between functions must also be examined.

The relationship between function of the type "function *f* has a buffer – function *g* has access to *f*'s buffer" is easily described through a graph where each node represents a function and the arcs between nodes represent an access from on function to another's internal buffer (Figure 3). This graph is built during the analysis of the code analysis of the phase 1.

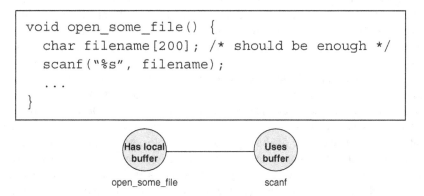

Fig. 3. Example of an unsafe buffer use involving two different functions (*top*). The *open_some_file* function contains the buffer and passes its address to the *scanf* function which uses this address ti fill the buffer without checking the size of the buffer. The relationship between the functions *open_some_file* and *scanf* is represented in the graph (*bottom*). Note that function *scanf* may itself call a third function passing the address to the buffer. However this would not change the analysis feasibility: the third function would be analyzed regarding its use of the buffer whose address it receives.

All functions that remain tagged as "suspect" are tested during phase 2. It is worth recalling that tests performed during phase two are defined based on information discovered during phase one. The first phase comprises seven tasks which are described next:

Task 1: Function Identification. This task consists on a recursive analysis of the target code for the identification of all its functions. If the observed software is an executable program, then the starting point of the analysis is the program entry point. If the observed software is a library module, such as a windows dll, then each exported function is a starting point. During the analysis, every time a call instruction is detected the target address is added to the list of functions to analyze, unless its address is in the list of functions already analyzed. This task is similar to a breadth-first tree traversal algorithm. The output is a list of all the functions accessible in the software being observed.

Task 2: Function Call-Called Relationship. This task is in fact performed at the same time as the previous one. The output is a graph establishing all relationships between calling and called functions.

Task 3: Stack Space Analysis. This task comprises the analysis of each function resulting from task 1 in order to determine the existence of buffer in the stack space of each function. There are two main clues that can be used to infer the existence of such buffers. One is the size and layout of the stack storage space (how many variables and what size they are), and the second is the type of address mode used by the instructions that deal with stack locations.

The stack size can be directly discovered in the function preamble. Following the example of the IA32 architecture, the number of bytes that is subtracted to the *esp* register is the amount of memory used to store the function local variables. Although there can be slight variations according to the processor or programming model used (e.g., for 16 bits programs the register is *sp* instead of *esp*), this value can always be discovered.

The layout of local stack storage space offers information on how many distinct variables are there and what is the size of each one. This information can be found by analyzing the code of the function and collecting all references to locations having the register *ebp* as base address and using a negative offset (once again there are slight variations in practice but the method remains applicable). As mentioned, there are two important characteristics that can be used to infer if a given stack-resident variable is a buffer: its size and the address mode used in the instructions which reference it. Buffers usually have a large size, and the instructions that refer to it usually use an base-indexed address mode (we present an example later on that illustrates this type of analysis).

Task 4: Function Parameter Analysis. This task is aimed at the identification of the type of the function parameters of potentially suspect functions. We are specifically interested in pointer type parameters. If a function does not have a local buffer but receives a pointer as parameter, then it is possible that the pointer points to the address of a buffer belonging to the calling function. It is worth mentioning that it is possible to infer if a given parameter is a pointer through the analysis of the instructions that use its value. All parameter are referred through positive offsets based on the register *ebp* (other processors use other registers but the method is the same). If a value obtained from such a location is later used as (part of) the destination address in a *mov* instruction, then it is probably a pointer.

Task 5: Local Buffer Use. This task applies to all functions found earlier that have a local buffer. During this task the code of the function is analyzed to determine the existence of instruction patterns related to the filling of the buffer with data or the calling of another function using the address of the buffer as a parameter. The first case can be detected through the occurrences of instructions patterns such as loops containing *mov* instructions, or special prefixed *mov* instructions such as *rep movs*. The calling of a function passing a pointer to the local buffer can be easily detected through the occurrence of load effective address instruction using the buffer followed by a push instruction before the call instruction (e.g., *lea reg, [ebp-offset]*, *push reg*, *call addr*).

Task 6: External Buffer Filling Detection. This task applies to the functions that receive a pointer as parameter. During this task the code of the function is analyzed to determine the existence of instruction patterns related to the filling of a buffer using that pointer or the calling of another function using the pointer (this process is similar to task 5).

Task 7: Function Classification. This task is responsible for the classification of functions as suspect or clear. All functions are initially tagged as clear. The rules that cause a function to be tagged as suspect are the following:

- **Rule 1**: A function that has a local buffer and fills it without checking the limits using a constant value equal to the size of the buffer is tagged as suspect.
- **Rule 2**: A function that receives a pointer and uses it as destination of a buffer-filling operation is tagged as suspect.
- **Rule 3**: A function that receives a pointer and uses it as parameter to another suspect function is tagged as suspect.
- **Rule 4**: A function that has a local buffer and passes its address to another function already tagged as suspect is tagged as suspect as well.
- **Rule 5**: All functions that reside outside the software under observation and receive a pointer as parameter are automatically tagged as suspect (calls to these functions are discovered during task 5).

It is worth mentioning that all functions that do not call other functions, or do not supply pointers to the called functions, must be processed before the others.

4.2   Example of Information Extraction from Low-Level Code

In order to illustrate the kind of analysis that is performed within each function, and to exemplify how the relevant information can be extracted from low-level instructions, we present a low-level code example. Figure 4 presents the instruction sequence of a given function. From the analysis of its instruction sequence we can immediately discover the following: there is a local variable with size 200; this variable is most likely a buffer due to its size and the kind of address mode used to access it (*mov [ebp+edx-204]*, …); there is a loop which copies bytes into the buffer; the loop is controlled by the value 200 (probably the size of the buffer); the first parameter (*ebp+12*) appears to be used as the number of bytes to place in the buffer; the second parameter (*ebp+8*) is used as source of the bytes to place in the buffer.

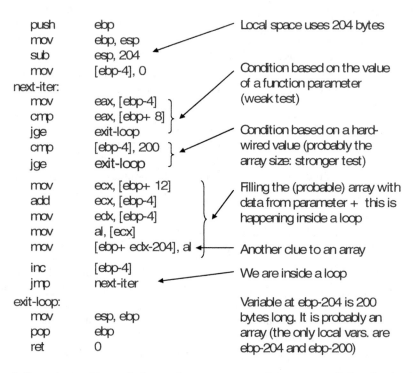

```
        push      ebp                        Local space uses 204 bytes
        mov       ebp, esp
        sub       esp, 204
        mov       [ebp-4], 0                 Condition based on the value
next-iter:                                   of a function parameter
        mov       eax, [ebp-4]  ⎫           (weak test)
        cmp       eax, [ebp+ 8] ⎬
        jge       exit-loop     ⎭            Condition based on a hard-
        cmp       [ebp-4], 200  ⎫            wired value (probably the
        jge       exit-loop     ⎭            array size: stronger test)

        mov       ecx, [ebp+ 12] ⎫          Filling the (probable) array with
        add       ecx, [ebp-4]   ⎪          data from parameter + this is
        mov       edx, [ebp-4]   ⎬          happening inside a loop
        mov       al, [ecx]      ⎪
        mov       [ebp+ edx-204], al ⎭      Another clue to an array

        inc       [ebp-4]                    We are inside a loop
        jmp       next-iter
exit-loop:                                   Variable at ebp-204 is 200
        mov       esp, ebp                   bytes long. It is probably an
        pop       ebp                        array (the only local vars. are
        ret       0                          ebp-204 and ebp-200)
```

Fig. 4. Example: machine-code instruction sequence pertaining to a small function having a local buffer. The existence of the buffer is deduced from the analysis of the stack layout and the references to it in the code. The function has a portion of code that fills the buffer. The analysis of that portion of code provides clues on the existence of buffer boundary checking.

From there we can conclude the relevant facts used as basis for the proposed methodology:

- There is a variable of size 200 in the stack that is very likely a buffer.
- The (possible) buffer is being filled with values supplied from a source outside the function. This has two consequences: a) it increases the probability that the large variable is indeed a buffer, and b) the content stored in the buffer is not determined by the function itself as its source is supplied as a parameter. If nothing else were known about this function it would be tagged as suspect.
- The loop where the buffer is filled is explicitly controlled by a hard-wired value which is equal to the size of the buffer. This leads to the conclusion that this function is not vulnerable to stack smashing and, as such, the function remains tagged as clear.

In the example presented above one important fact discovered during the function analysis that is relevant to the testing phase is the following: if more than 208 bytes are stored in the buffer, a stack smashing will occur (see Figure 5). Should the function not check explicitly for the buffer limit, then it would have been tagged as suspect and tested during phase two. In such case the values supplied as parameters would include a value larger than 208 (first parameter) and a pointer to a sequence of more that 208 bytes (second parameter). Should the buffer overflow occur and smash the stack, the function would most likely crash because the return address was overwritten.

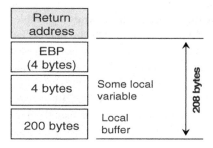

Fig. 5. Stack layout of the example presented above. If more than 208 bytes are stored in the local buffer, the return address is overwritten. Thus, the testing of the example function during step 2 must necessarily include the calling of the function supplying at least 209 bytes to store in the buffer.

4.3   Phase 2 – False Positive Elimination

The second step of our methodology is devoted to the test of the functions tagged as suspect in the previous step. This step is required because the analysis process conducted during step one of our methodology does not guarantee that all functions tagged as suspect are indeed vulnerable. In fact, the general problem of using one program to decide the correctness of another program has been shown to have no general solution. Our approach remedies this by taking a pessimistic approach during the first step and tagging every function that may be vulnerable to stack smashing as suspect. During the second step the false positives are eliminated through specific testing.

The testing process is conducted in a manner similar to traditional robustness testing such as [16]: the observed software is executed and supplied with specially tailored inputs to attempt to uncover its internal vulnerabilities. However, there are two very important differences when comparing our method to traditional robustness testing. The first is in the fact that not every function is tested. Only those that were tagged as suspect are subjected to testing. This reduces the time needed for the experiments. The second difference is in the fact that the values supplied to the functions being tested are based on the information discovered during the first step, and not just based on the function parameter input domains as in robustness testing. This increases the likelihood of confirming the vulnerability of the function being tested and again reduces the time needed for the testing experiments.

As mentioned, there are basically two formats that the observed software can take: a self-contained executable program, and a library such as a dynamic link library. The testing process of functions belonging to libraries is a straightforward process. Each suspect function is directly invoked from the experimental setup. The values supplied as parameters are those resulting from the information discovered in the step one.

It is possible that some of the functions being tested are dependent on (call) functions external to the software being tested (e.g. operating system API functions). External functions which are supplied with the address of buffers are relevant for the testing process (recall that such situations are identified during step one). The fact that the address of an internal buffer is supplied to the exterior of the software (which is not directly controlled by the programmer) exposes a possible vulnerability. To verify if that vulnerability is in fact present these calls are intercepted and the memory at the address supplied is filled in an attempt to cause a stack smashing.

The testing of a self-contained program is similar to the testing of a library. The most relevant difference is the fact that the testing process is entirely guided through the interception of external functions being called from within the program. This is due to the fact that a self-contained program has exactly one entry point and usually do not export functions as a library does.

5 Methodology Portability Discussion

Our methodology is based on the automatic analysis of executable code to extract information on possible vulnerabilities. This information is present in the very structure of the executable (e.g., the stack layout) and in the patterns of the instruction sequences. Software programs are in fact a kind of standardized constructs: programs are organized in code, data, heap and stack; the code is organized in modules which call another modules; each module stores local variables in stack, and so on. Although specific details may vary, a structured programming model is used on all platform architectures (see example in Figure 6). Thus, our methodology is in itself portable as there is nothing in the methodology tied to a specific platform architecture or program model (the IA32 architecture was used in the paper only as an example).

Obviously, different processors have different instructions sets. However, the sole consequence of that fact is that there must be a specific implementation of the methodology for each intended target processor. This should not be viewed as a methodology impairment since one implementation able to analyze software for several platforms is not necessarily better than several implementations targeting one platform each. Different implementations differ mostly on the meaning assigned to each processor instruction. The algorithm of the methodology remains the same. Most platform architectures and processors share a common programming paradigm. More specifically, the same source-code constructs tend to result in similar instruction patterns. Obviously, there are differences in the instruction patterns, especially in what concerns optimization strategies. However, the analysis process of phase one of our methodology is easily adapted to the particular characteristics of different processors.

Code example	Tipical instruction sequence	Explanation
procedure entry	ldgp $gp,0,($27)	lod procedure value
	lda $sp,-*framesize*($sp)	Framesize for local vars and saved regs
tasks:	stq $ra,0($sp)	Save return address
set frame stack and	stq $fp,8,($sp)	save fp
save registers	mov $sp,$fp	sets framepointer

Fig. 6. Example of Alpha processor instructions related to the entry-point and stack frame initialization. As in the case of the IA32 architecture, the code is structured in a predictable way and all the information necessary to our methodology can be extracted from the code.

6 Application Scenarios

Several application scenarios can benefit from this methodology: evaluation of existing systems for which no source-code is available, COTS evaluation in COTS-based software development, and definition of attackloads for security testing and benchmarking.

The evaluation of software systems from a security perspective is a kind of evaluation that is severely impaired when no source code is available. Many security flaws are based on poor programming techniques which traditionally can only be detected by analyzing the system source code. The methodology presented in this paper enables the detection of security flaws based on buffer overflow exploitation without requiring source code.

Given the current industry trend of building computer systems through the integration of general-purpose components off-the-shelf, there is an increasing risk of system failure due to the integration of vulnerable components. Usually, COTS source code is not available. This foils most attempts to effectively test these components from a security perspective. As our methodology does not depend on source code availability (or even the availability of any kind of information about the internal architecture of the target software), it presents a valuable tool in the effort of testing general purpose COTS software components.

One recent effort in the dependability research community is the definition of dependability benchmarks [18, 19, 20, 21]. Given the infeasibility of producing 100% error free systems, dependability benchmarks are assuming a crucial role in the evaluation of systems. A simplified definition of dependability benchmarking can be stated as being an extension to the traditional benchmarks concept where new components are added. These new components are related to the observation of the target system in the presence of faults: this means the use of fault injection techniques and the use of faultloads. The definition of faultloads is perhaps the most complex task when proposing a new benchmark. One particular class of faults that is very hard to define and specify in a faultload is the malicious fault (i.e., the definition of an attackload). These are not faults in the traditional sense; instead they are user actions specifically aimed at the exploitation of system vulnerabilities. One reason that makes malicious faults hard to specify is that the vulnerabilities exploited by attacks are usually unknown (except for the attacker) and are very dependent on the system being attacked. In this scenario, our methodology can be very useful in the detection of weaknesses in commonly used software libraries. The weaknesses discovered can then be used to specify an attack which would be relevant to the class of systems that employs the observed software library.

Another application scenario that should be taken into account is the possible use of this methodology by malicious hackers to discover weaknesses to exploit. Thus, it is necessary to consider if the existence of such a methodology provides more advantages than disadvantages. This question is in fact a particularization of the more general issue of open software vs. close software: is the open software more secure because there are more programmers with access to the source code that discover more bugs, or is the closed software more secure because the malicious hackers do not have access to the source code. Although both positions are defensible (with perhaps more supporters to the open-source), there is a general consensus that the advantages of the source code availability ("more eyes make all bugs shallow" [22]) overcomes its disadvantages (see [23]). Thus, it is our opinion that the existence of our methodology does not increase the risk to software systems while at the same time it provides a valuable means for detecting vulnerabilities in third-party software (this is in fact the same reasoning behind the existence of tools to test network security flaws).

7 Conclusion

We presented a methodology for the automated discovery of vulnerabilities leading to buffer overrun and stack smashing attacks. As nearly half of all the successful computer-system attacks are based on the exploitation of buffer overrun and stack smashing vulnerabilities, a technique to detect these vulnerabilities is particularly useful. The methodology does not depend on the availability of source code nor in any prior knowledge about known weak library functions. This enables the application of the methodology in the COTS integration scenario, which is currently a major industry trend. The benefits deriving from this methodology are several: the integration of vulnerable software can be avoided; if no other choice is available, wrappers can be built to remove the vulnerabilities. The definition of attackloads can also be based on the vulnerabilities discovered through the application of this methodology. Given the increasing number of vulnerabilities being exploited to attack computer systems, the definition of attackloads is a problem of high relevance.

Another important aspect of this methodology is the fact that it is not tied to a particular processor and can be implemented for any given platform architecture.

References

1. FrSIRT – French Security Incident Response Team (available online at http://www.frsirt. com/ exploits/)
2. M. Eichin, J. Rochlis, "With microscope and tweezers: An analysis of the internet virus of November 1988", Proceedings of the 1989 IEEE Computer Society Symposium on Security and Privacy.

3. Microsoft Security Bulletin MS04-028, "Buffer Overrun in JPEG Processing (GDI+) Could Allow Code Execution", available at www.microsoft.com/technet/security/bulletin/ MS04-028.mspx.

4. T. Chiueh, F. Hsu, "RAD: A Compile Time Solution for Buffer Overflow Attacks", Proceedings of the 21st IEEE International Conference on Distributed Computing Systems (ICDCS), April 2001

5. C. Cowan, et al, "StackGuard: Automatic Detection and Prevention of Buffer-overrun Attacks", Proceedings of the 7th USENIX Security Symposium, January 1998

6. D. Larochelle, D. Evans, "Statically Detecting Likely Buffer Overflow Vulnerabilities", Proceedings of the 2001 USENIX Security Symposium, Washington, D. C., August 13-17, 2001

7. A. Baratloo, N. Singh, T. Tsai, "Transparent Run-Time Defense Against Stack Smashing Attacks", Proceedings of the 2000 USENIX Annual Technical Conference, San Diego, California, USA, June 18-23, 2000.

8. A. Ghosh, T. O'Connor, "Analyzing Programs for Vulnerability to Buffer Overrun Attacks", Technical Report, Reliable Software Technologies, January 1998

9. Joey__ (Nishad Herath), "Advanced Windows NT Security", The Black Hat Briefings'00, April 2000, Singapore.

10. J Newsome, D Song, "Dynamic Taint Analysis for Automatic Detection, Analysis, and Signature Generation of Exploits on Commodity Software", Proceedings of the 12th Annual Network and Distributed System Security Symposium – NDSS-05, February 2005.

11. M. Christodorescu, S. Jha, "Static Analysis of Executables to Detect Malicious Patterns", Proceedings of the 12th USENIX Security Symposium, August 2003

12. T. Gillette, "A Unique Examination of the Buffer Overflow Condition", MsC Thesis, 2002, College of Engineering of the Florida Institute of Technology.

13. Aleph One, "Smashing the stack for fun and profit", Phrack Magazine, 49-14, 1998

14. J. Durães, H. Madeira, "Emulation of Software Faults by Educated Mutations at Machine-Code Level", Proceedings of the Thirteenth IEEE International Symposium on Software Reliability Engineering, ISSRE'02, November 2002, Annapolis MD, USA.

15. J. Durães, H. Madeira, "Definition of Software Fault Emulation Operators: a Field Data Study", in Proceedings of International Conference on Dependable Systems and Networks - DSN2003, San Francisco, 2003 (IEEE William Carter Award for the best student paper).

16. P. Koopman et al, "Comparing Operating Systems using Robustness Benchmarks", Proceedings of the 16th International Symposium on Reliable Distributed Systems, SRDS-16, 1997

17. J. C. Fabre, M. Rodríguez, J. Arlat, F. Salles, and J. M. Sizun, "Bulding Dependable COTS Microkernel-based Systems using MAFALDA", in Proceedings of the 2000 Pacific Rim International Symposium on Dependable Computing - PRDC'00, 2000, pp. 85-92

18. A. Brown, D. Patterson, "Towards availability benchmark: a case study of software RAID systems", Proceedings of 2000 USENIX Annual Technical Conference, San Diego, California, USA, June 18-23, 2000, pp 263-276.

19. K. Kanoun, J. Arlat, D. Costa, M. Dal Cin, P. Gil, J-C. Laprie, H. Madeira, N. Suri, "DBench: Dependability Benchmarking", Supplement of International Conference on Dependable Systems and Networks, DSN-2001, Göteborg, Sweden, 2001

20. Marco Vieira, and Henrique Madeira, "A Dependability Benchmark for OLTP Application Environments", 29th International Converence on Very Large Databases, VLDB 2003, Berlim, Germany, Sept. 9-12, 2003

21. João Durães, Marco Vieira, and Henrique Madeira, "Dependability Benchmarking of Web-Servers", The 23rd International Conference of Computer Safety, Reliability and Security, SAFECOMP 2004, Potsdam, Germany, September 21-24, 2004.

22. E. Raymond, "The Cathedral and the Bazaar", 1998, available at http://tuxedo.org/~esr/writings/cathedral-bazaar/

23. R. Anderson, "Security in Open versus Closed Systems – the dance of Boltzmann, coarse and moore", Proceedings of the Open Source Software Economics, Law and Policy, Toulouse, France, June 20-21, 2002, available at http://www.ftp.cl.cam.ac.uk/ftp/users/rja14/toulouse.pdf

Quantitative Evaluation of Distributed Algorithms Using the Neko Framework: The NekoStat Extension

Lorenzo Falai[1], Andrea Bondavalli[1], and Felicita Di Giandomenico[2]

[1] DSI - Università di Firenze, Viale Morgagni 65, I-50134 Firenze, Italy
{lorenzo.falai, bondavalli}@unifi.it
[2] ISTI CNR, Via Moruzzi 1, I-56124 Pisa, Italy
digiandomenico@isti.cnr.it

Abstract. In this paper we present NekoStat, an extension of the Neko tool. Neko is a Java framework and a communication platform that permits rapid prototyping of distributed applications; it provides tools to organize the applications using a layered architecture, with the network(s) at the bottom of the architecture. Neko is also a communication platform that allows sending and receiving of generic Java objects. Distributed systems realized within the Neko framework can be exercised both on real networks and on simulated ones, without changes in the application code. We constructed an extension to plain Neko, called NekoStat; it permits attainment of quantitative evaluations of distributed systems. In the paper we describe this extension; we motivate the development of NekoStat, we describe the design and finally we illustrate its usage through a case study, which highlights the usefulness of NekoStat.

1 Introduction

The quantitative evaluation of performance and of dependability-related attributes is an important activity of *fault forecasting* ([1]), since it aims at probabilistically estimating the adequacy of a system with respect to the requirements given in its specification. Quantitative system assessment can be performed using several approaches, generally classified into three categories: *analytic*, *simulative* and *experimental*. Each of these approaches shows different peculiarities, which determine the suitableness of the method for the analysis of a specific system aspect. The most appropriate method for quantitative assessment depends upon the complexity of the system, the development stage of the system, the specific aspects to be studied, the attributes to be evaluated, the accuracy required, and the resources available for the study. Analytic and simulative approaches are generally cheap for manufacturers and have proven to be useful and versatile in all the phases of the system life cycle. They are typically based on a parametric model of the analyzed system and on a set of assumptions concerning the behavior of the system and/or of the system environment. Analytic approaches are highly efficient, but the accuracy of the obtained results is strongly dependent

C.A. Maziero et al. (Eds.): LADC 2005, LNCS 3747, pp. 35–51, 2005.

upon the accuracy of the values assigned to the model parameters and on how realistic the assumptions the system model is based on are. The simulative approach is one of the most commonly used approaches for quantitative evaluation in practice, especially for highly complex systems, for which analytical solutions are generally precluded; however, it tends to be generally more expensive. As for the analytic approach, the accuracy of the obtained evaluation depends on the assumptions of the analyzed system as well as on the behavior of the environment, and on the simulation parameters; however, it is superior to analytic models in capturing relevant phenomena through more realistic representations (e.g., to overcome the exponential distribution for events occurrences, which is usually implied by the analytic solution). Experimental measurement is an attractive option for assessing an existing system or prototype. This method allows monitoring the real execution of a system to obtain highly accurate measurements of the metrics of interest. However, it may turn out to be quite expensive, e.g., when the interest is in very rare events, and moreover the obtained results are very difficult to generalize.

The largeness and complexity of dependability critical systems, together with the necessity of continuous verification and validation activities during all the design and development stages in order to promptly identify deviations from the requirements and critical bottleneck points, call for a composite V&V (verification and validation) framework, where the synergies and complementarities among several evaluation methods can be fruitfully exploited. Comparison of results for a certain indicator obtained through the application of two alternative methods allows *cross-validation* of both. Feeding a system model with parameter values derived through experimental measurement is a central example of *cross-fertilization* among different methods.

A high proliferation of automatic tools supporting a variety of quantitative evaluation methods has been reached till now, and research in this direction is always in progress. In recent years, special attention is being devoted to the analysis of distributed protocols. Distributed protocols are highly employed as "basic building blocks" providing basic services, on top of which distributed applications run. To exemplify, communication protocols (reliable broadcasts/multicast), consensus algorithms, clock synchronization, failure detectors to diagnose faulty components, are among the most typical representatives in this category. Distributed protocols have been traditionally analyzed and validated in terms of qualitative properties, such as "termination", "validity", "integrity", and "agreement". Most of these properties are specified in terms of "eventual behavior" (e.g., "sooner or later a faulty processor is identified as faulty" or "a decision will be made"). To verify such properties is appropriate and important for theoretical correctness, whereas in real world contexts, where time aspects are relevant, assessment of quantitative metrics is necessary and required (e.g., when guarantees are required on the ability to perform a certain number of runs of a certain consensus protocol in a fixed time interval). It is therefore paramount to identify metrics useful to specify dependability, performance and quality of service (QoS) requirements of distributed protocols, and to define methods and tools to analyze and validate protocols against these QoS metrics.

In this stream of methodologies and tools to contribute to the V&V of distributed algorithms, a framework has been developed, called Neko, which consists of a simple communication platform that allows to both simulate a distributed algorithm and execute it on a real network, using the same implementation for the algorithm ([2,3]). Coupling both simulative and experimental execution inside the same environment turns out to be very convenient, for a number of reasons: the easiness of usage for the user (who has to learn just one tool); the efficiency due to the inherent integration of the methods; and the increased confidence to analyze always the same system, since the same protocol implementation is used in all types of analysis.

However, Neko permits only to collect *traces of execution*; it does not include support to collect and manage events so as to perform on-line quantitative evaluations, in parallel with the algorithm execution. The contribution of the work described in this paper is in this direction, and we present the NekoStat extension to Neko, through which it is possible to perform simple and powerful quantitative analysis of distributed algorithms, using simulative and experimental approaches. Following the same philosophy of Neko, NekoStat has the ability to perform quantitative evaluations adopting both the simulative and experimental approaches. The main difference between these two kinds of analysis is that in simulations we can make on-line evaluations, whereas in real experiments the quantitative evaluation is performed only at the termination of the distributed system execution, after all the data have been collected.

The rest of the paper is organized as follows. In Section 2 we describe the basic architecture and usage of the Neko framework. Section 3 is devoted to the description of the newly developed NekoStat package. Section 4 illustrates the usage of NekoStat on a case study. Finally, conclusions and indications of further developments are drawn in Section 5.

2   The Neko Framework

Neko ([3]) is a simple but powerful framework that permits the definition and the analysis of distributed algorithms, showing the attracting feature that the same Neko-based implementation of an algorithm can be used for both simulations and experiments on a real network. This way, the development and testing phases are shorter than with the traditionally used approach, in which these two implementations are made at different times and using different languages (for example, SIMULA for simulations and C++ for prototype implementation). Neko was built at the Distributed Systems Lab of the EPFL in Lausanne, it is written in Java to assure high portability and has been deliberately kept simple, extensible and easy to use.

The architecture of Neko can be divided in three main components (see Figure 1): *applications*, *NekoProcesses* and *networks*.

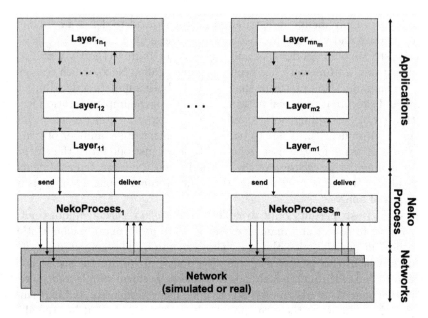

Fig. 1. Typical architecture of a Neko-based distributed application

Applications are built following a hierarchical structure based on multiple levels (called *Layers*). Layers communicate using two predefined primitives for message passing: send transports a message from a layer to the level below, and deliver transports a message in the opposite direction, from one layer to the next upper level. Note that messages in Neko may contain generic Java objects. Then, a typical Neko-based distributed application is composed by a set of m processes, numbered $1, ..., m$, communicating through a message passing interface: a *sender process* inserts, through the asynchronous primitive send, a new message in the network, and the network delivers the message to the *receiver process* through the deliver primitive.

There are two types of layers: *active* and *passive*. Active layers have an associated thread (and thus an application logic independent from the other layers of the stack), whereas passive layers can only react to send and deliver of messages.

Every process composing a distributed application has associated to it a NekoProcess, which maintains data common to all the layers (e.g. the address of the local host) and it contains the implementation of some services of general utility (e.g. forwarding of messages to the right network).

The Neko communication platform is a *white box*: the developer can use a network available on Neko or he/she can define new network types. Different networks can be used in parallel, and this allows the exchange of different types of message using different networks (e.g. it is possible to use a TCP connection for some message types and UDP datagram service for others).

Neko *networks* are the lowest level of the architecture of a Neko application. As already mentioned, an implementation of a distributed algorithm can run on top of a real network, as well as on a simulated network, without changing any line of code. In fact, two types of networks are supported by Neko: *real* and *simulated* networks. Real networks are built from Java sockets, or using external libraries for proprietary networks. Sending and receiving of a NekoMessage in real networks is based on *serialization*, an operation that permits message representation during the message passing between processes. The Neko communication platform provides several predefined simulated networks. However, integration of a new network type can be easily performed. This operation requires the definition by the developer of a new model for the network, to be expressed using a new **NekoNetwork** subclass (and defining the associated **send** and **deliver** methods). It is therefore possible to use in a Neko application a proprietary network, or to define a new kind of simulated network (e.g. a network in which delays and losses follow specific distributions).

A Neko application can be configured through a configuration file, containing information to set up all the involved processes. Then, bootstrapping a Neko application is different for a simulation and a distributed execution. In real executions there is an asymmetry between different processes: there is a *master process*, that coordinates the execution, and $m - 1$ *slave processes*. The master is the process that provides the configuration file to the slaves. The m processes run on m Java Virtual Machines, usually executing on m different hosts, which communicate using the communication platform of the framework. Simulation startup is simpler; there are m processes, in execution as different threads of one Java Virtual Machine. A complete description of Neko can be found in [2].

Neko can be used to make *dynamic testing* of distributed algorithms and systems. It permits testing of a system searching for *qualitative properties*; the tool is equipped with supports to obtain *execution traces*, both on simulated and on real environments. The potentialities of the Neko tool in the rapid prototyping of distributed algorithms are thus evident: the possibility to use simulated networks permits the analysis of the algorithm in different conditions (variable transmission delays, different probabilities of message losses, network congestion, ...) and, after that, it is possible to test the algorithm in real environments. Neko is thus very useful and versatile to test new algorithms, or to compare already existing ones. Neko also allows performing *fault injection* experiments at the network level as well as at the level of communications between layers, and thus it can be used to study the behavior of the analyzed algorithm with respect to specific injected faults or exceptional situations.

Although possessing the attractive features exposed so far, the Neko framework lacks any support to quantitative assessments. In fact, the kind of analyses supported by Neko is directed to assess *qualitative properties* (so-called "on/off" properties) of distributed systems, following a dynamic verification approach. Thus quantitative measurements can be obtained only *off-line*, through awkward manipulation of the logs collected, i.e. history of the distributed execution. There is no embedded support to help the quantitative evaluations. Such a quantitative analysis can be made possible through re-definition of supports inside

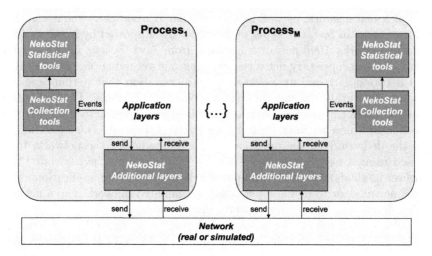

Fig. 2. High level view of typical session of analysis of a distributed system made with NekoStat

any single application, or through setting up proper filters able to interpret the traces of the executions collected in the log file.

Therefore, to permit assessment of *quantitative properties* - namely, dependability and performance metrics - we devised, designed and constructed an extension to standard Neko framework.

3   The NekoStat Package

NekoStat extends the V&V analysis features of Neko in the direction of a *statistical dynamic evaluation* of a system, both on simulated and real execution. In Figure 2 a high level view of the structure of a session of analysis using the NekoStat extension is depicted.

One of the basic ideas of Neko was to define a framework in which the developer can use the same implementation of a distributed algorithm, both for simulations and real experiments. We wanted to retain this appealing feature in the NekoStat design: the usage of the provided tools is similar both in simulations and in real experiments.

The quantitative analysis of distributed systems in NekoStat is approached through two sets of tools: the first set is composed of tools to *collect events*, whereas the second one includes tools for the *analysis of measurements*.

Using NekoStat to obtain an assessment of relevant metrics of a distributed system is simple. First of all, it is necessary to implement the distributed system/algorithm, using the tools available in the Neko framework (Java language). Then, in order to apply the tools available in the NekoStat extension, the following simple modifications have to be performed to the application code:

1. define the interesting *events* and introduce calls to the log(Event) method, of a special predefined logger class (StatLogger) in the points of the source code where the event happens;
2. implement a StatHandler, a class containing the methods to manage the collected *distributed events* and to transform them into *quantities*.

Figures 3 and 4 show the architectures of a typical analysis session obtained using NekoStat, for simulations and real executions respectively. As depicted in the Figures, a part of the NekoStat support classes must be defined by the user: it can be used both for simulative and experimental evaluations. The definition of the StatHandler and of the Quantities is dependent on the *analyzed distributed application* and on the *interesting metrics*.

The NekoStat functionalities, and the related components implementing them, can be subdivided in two sets: *mathematical functionalities*, that handle the numerical quantities, and *analysis functionalities*, that collect and analyze distributed events. The implementation of the mathematical functionalities is the same both for simulation and real executions, whereas analysis supports are internally different, still with a common interface. However, differences in the internal structure are hidden to the developer, so the same code for the analysis can be reused in experiments and simulations without changes.

Although the interface between NekoStat and the application layers is the same for simulative and experimental analysis, the evolution of the analysis is different. A simulation starts with the creation of the Neko and NekoStat parts of the architecture depicted in Figure 3. The application layers and the user-

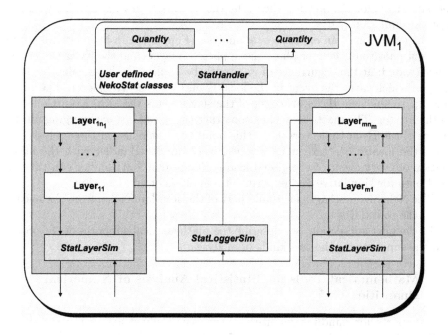

Fig. 3. Example of architecture for simulative evaluation

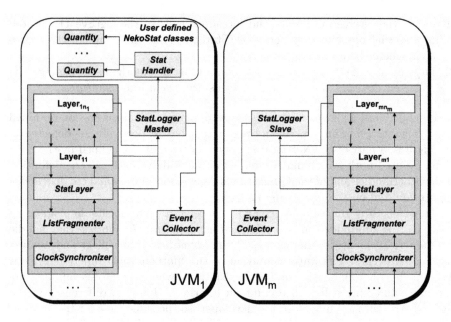

Fig. 4. Example of architecture for experimental evaluation

defined StatHandler are immediately activated; at the occurrence of an event, the application layer calls the StatLogger, which calls the StatHandler with the information on the event. The StatHandler can use this data to obtain the measurements for the metrics; the evaluation is thus *on-line*, in parallel with the system simulation.

The evolution of an experimental analysis (Figure 4) can be subdivided into different phases. In the first phase the application layers and the EventCollectors are activated; at the occurrence of an event, the application layer calls the Stat-Logger, which saves the event in the local EventCollector. At the termination of the experiment run, the StatLogger of the slaves sends the local EventCollector to the master. The master can thus construct the *global history*, merging all the events of the EventCollector(s). At this point the last phase of the analysis can start: the master StatLogger calls repetitively the StatHandler with the information of every event of the global history. The same StatHandler can thus be used both for simulative and experimental analysis of an algorithm.

The NekoStat package is actually part of the standard Neko, from the release 0.9 of the tool ([4]).

In the next subsections we describe the mathematical supports of NekoStat, and the supports for the two kinds of analyses.

3.1 Mathematical Tools for Statistical Analysis of Numerical Quantities

The supports for handling quantities have been defined as expansions of the *Colt* mathematical library for Java ([5]). The Colt library was developed at

Cern, with the purpose of offering a powerful and fast mathematical library for Java (in place of default tools available in the java.Math package).

The classes defined for statistical analysis are composed of a *values container* and *methods* to obtain statistical parameters. We built different classes to handle the *metrics* (quantity classes); such classes are characterized by different memory usages and different statistical methods usable on collected data.

Every class has:

- an **add** method that permits to insert a new measurement to the collection;
- methods to obtain statistical parameters of the quantity, such as the dimension of the collection, the sample mean, the median value, the sample standard Deviation, the minimum and maximum values measured,...

The complete list of the information obtainable from a quantity can be found in the Colt library documentation ([6]).

There are also some other methods available to all the quantity classes: initial transient elimination, export of main statistical parameters on file, definition and handling of stop conditions for simulation.

The stop condition is a boolean variable, whose value is true when the already observed values give enough confidence on the quantity under analysis; the value of the stop condition can be observed during the simulation to decide if the confidence on all the analyzed quantities is satisfactory enough.

We defined two stop conditions:

Stop after N values: after N collected measures of the quantity, the stop condition becomes *true*;

Stop on confidence interval: the stop condition becomes *true* when the confidence interval of the mean of the quantity, of level $(1 - \alpha)$, is less than β percent of the mean. After n measures, let $\widehat{x}$ be the estimated mean, S^2 be the estimated variance of the quantity; we thus decide when enough data have been collected using the expression:

$$t_{n-1,1-\frac{\alpha}{2}} \sqrt{\frac{S^2}{n}} \leq \widehat{x} \frac{\beta}{(1 - \beta)}$$

See [7] to obtains further information about the expression above.

We recall that the use of a stop condition is applicable only for on-line analysis in simulations; in the experimental evaluations the quantities are evaluated at the termination of the distributed execution, using all the collected events.

A more detailed description of the classes and their usage is in [8].

3.2 Analysis Supports

The supports to quantitative analysis address different objectives, in accordance with the analysis method they are applied to. In more detail:

- For **simulations** we defined means for handling the quantitative analysis process. It is possible to define *stop conditions* for the simulation, essentially based on the *number* of evaluations obtained for a quantity or on the *accuracy* of the obtained results (this last is usually based on the *interval of confidence* for the mean of the quantity).

– For **real executions** we defined supports for *events collection* and for *clocks synchronization*.

The events collection supports have been developed so as to interfere as little as possible with the distributed execution: for example, the support that performs the local process event collection uses a specially defined *SparseArrayList*, built with an internal logic to minimize the overhead of NekoStat monitoring components [1].

Processes composing a Neko application executed on a real network have a logical clock, whose origin is at the start of the process; the Neko clock has granularity of 1 *msec*. In a real execution environment, we often need a *global clock* to evaluate and assess the temporal metrics, generally based on events that occur in different processes of the system. The user can employ the most appropriate policy to synchronize the clock of the hosts with the *real time* (e.g. using the Network Time Protocol [9], as in the example reported in the next Section). At the beginning of the experimental evaluation, the Neko logical clocks are also synchronized, using a simple master-slave approach, in which the real time corresponding to the start of the Neko master process is forwarded to all slaves: in this way, the slaves set the origin of time of the local logical clock to the real time corresponding to the origin of time of the master process (so as to create a unique, logical, global clock). This approach has the advantage of providing good usability and good precision of the clocks in different contexts, from LAN to WAN environments ([10]).

– Both for **simulations** and for **real executions** we defined supports to handle the *start* and *stop* of the analysis phases; to perform this, we defined appropriate layers, hidden to the NekoStat user.

4   Example of Usage

To show the usage of NekoStat and to point out its appealing functionalities to support quantitative analysis of distributed systems, in this Section we describe in detail a case study analyzed using the new tool. It is an experiment that we made on a Wide Area Network connection, between Italy and Japan, to evaluate and fairly compare the Quality of Service of a large family of adaptive failure detectors.

The ability to detect component failures is a qualifying feature of distributed systems. The *unreliable failure detectors*, introduced by Chandra and Toueg in [11], are one of the most used approaches to failure detection. The failure detectors are *distributed oracles* that provide *hints* (unreliable information) on the failures of the system. An asynchronous system equipped with several kinds of

---

[1] In our measurements the mean time to insert a new Event into the special list was 4 *microseconds*, and the maximum value was around 1 *millisecond*: these are values compatible with typical NekoStat usage.

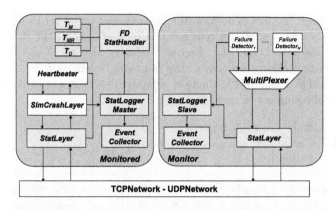

Fig. 5. Example of NekoStat architecture for a comparison of different failure detectors solutions through an experimental approach

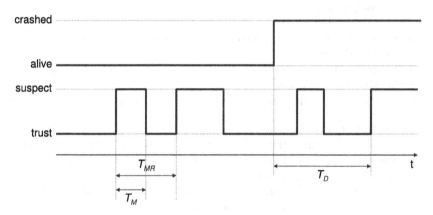

Fig. 6. Base metrics for the QoS evaluation of a failure detector

unreliable failure detectors is a very popular system model for distributed computations. In [11] eight classes of failure detectors have been formalized and classified according to the *logical properties* of *accuracy* and *completeness* of the information supplied by the failure detector. The qualitative classification proposed in [11] can however be inadequate. The Quality of Service (QoS) obtainable at the application level is related to the QoS of the failure detector used; especially for applications with temporal requirements we thus need *quantitative evaluation* of the QoS of the failure detector. For this reason we chose to quantitatively assess the QoS of a large family of adaptive failure detectors (FD) on a real WAN connection.

The architecture of our experiment is depicted in Figure 5. The distributed system used is composed of two Neko processes: *Monitored* and *Monitor*. The Monitored process periodically sends a new heartbeat message (from the Heartbeater layer), which is used by the failure detectors to establish whether the monitored process is alive. The Monitor process contains a set of different failure detector alternatives: 30 different calculation methods for the timeout of the failure detectors have been used. Figure 5 depicts also the NekoStat support classes that permit the attainment of the numerical evaluation.

As described in detail in [12], the QoS of failure detectors can be characterized through the following set of metrics:

- T_M, the mistake duration time;
- T_{MR}, the interval between successive mistakes;
- T_D, the detection time.

Figure 6 provides a graphical representation of these basic metrics.

Actually, this is a complete set: combining these metrics it is possible to obtain more sophisticated QoS indicators, which better represent how good a FD is in relation with the characteristics of the application using the FD.

The measurements for these metrics can be derived using the time interval between the following events:

- $StartSuspect_i$, the time necessary to the i-th failure detector to start suspecting the Monitored process;
- $EndSuspect_i$, the time necessary to the i-th failure detector to stop suspecting the Monitored process;
- $Crash$, the time to the crash of the Monitored process.

Calls to the $log(event)$ of the $StatLogger$ have been inserted in the points of the source code in which the events happen: for example, in the point of the source code corresponding to the receiving of a *fresh enough* heartbeat message, we introduced a call $statLogger.log(new\ Event("EndSuspect_i"))$. The evaluations are done by the $FDStatHandler$: it receives the events above described, and from these it extracts values for the metrics.

To obtain a simple, fair and complete evaluation of the QoS metrics we used two special layers. The first one is *MultiPlexer*, performing a simple task: it forwards every received message from the lower level to the upper layers, and it forwards also messages from the upper layers to the lower level. This layer permits feeding directly the different failure detectors, guaranteeing that they perceive identical network conditions, and thus it is the basis to fairly compare their QoS. The *SimCrashLayer* is instead a layer that injects a crash of the *Heartbeater* layer: it permits the evaluation of the behavior of the failure detectors in presence of a crash failure of the Monitored process (obtaining thus values for T_D).

We now consider an example of execution trace. Figure 7 depicts the trace, the distributed events and the obtained measures. In the Figure three layers are depicted: the Heartbeater, the Multiplexer, and FD_i, one of the failure detectors. The trace corresponds to a so-called *good period*: the Monitored process is up. The heartbeat messages hb_n and hb_{n+1} arrive to the Multiplexer, whereas hb_{n+2} is lost on the network. The Multiplexer immediately forwards the arrived heartbeats to the failure detector FD_i. In the bottom part of the Figure the

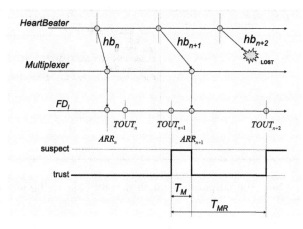

Fig. 7. Example of trace of the Failure Detector experiment

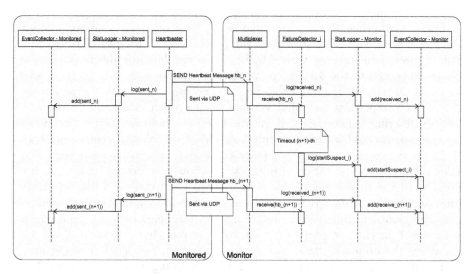

Fig. 8. UML Interaction diagram of the collection of the Events of the trace depicted in Figure 7

state of the failure detector FD_i is depicted: the state can be *suspect* or *trust*, and it can change in correspondence to a *timeout* or when a new heartbeat arrives. The measurements for the metrics T_M and T_{MR} can be obtained from the evolution of the state of the failure detector.

Figure 8 depicts the interaction diagram corresponding to this execution trace, expresses in the Unified Modeling Language (UML). Figure 8 represents what happens while the experiment is running: at the occurrence of the events, the application layers call $log(Event)$ of the local StatLogger, and the events are thus collected in the local EventCollector. When the experiment stops, NekoStat

tm01.data	tm01.result
2.0 7.0 7.0 3.0 1.0 3.0 980.0 4.0 1.0 1.0 [...]	Mean: 203.6723076923077 Min: 0.0 Max: 1206.0 StandardDeviation: 383.3223773727101 Measures: 650 Confidence Interval (95.0 %) : 29.468334142122792

Fig. 9. Example of NekoStat output files: results obtained for the T_M metric of a failure detector

rebuilds the *global distributed history*, from which it can extract the measurements under analysis.

The experimental evaluation of the QoS of the failure detectors was based on 13 different experiment runs, every one lasted for around 12 hours, on a WAN connection between Italy and Japan: the process Monitored ran in Italy (on a host connected to Internet with ADSL), while the Monitor ran in Japan (on a host connected to the JAIST[2] network). The different executions have been automatically run using a script that permits execution of a *batch of experiments*. This script also permits the definition*variables*, to which values can be assigned, possibly differing from one experiment run to others; however, in this case study, the experiment runs have been executed with identical values for the parameters.

The post-processing necessary to extract values for the metrics was negligible. NekoStat permits exportation of information about the metrics on files, in two formats: all the measurements for a metric can be stored on a file, and some other comprehensive data (like mean, standard deviation, and other user-defined indicators) can be exported on another file. An example of these output files is in Figure 9: in the left part of the Figure it is reported a portion of the file containing the obtained measurements for the metric T_M of a failure detector, whereas in the right part it is reported the file containing the comprehensive data of the same metric. The post-processing thus was only related to extract average values to summarize the QoS metrics for each failure detector with unique values. The possibility of extracting all the measurements obtained for a metric is useful also to study the distribution of values of the metric.

The *SimCrashLayer* and the *Multiplexer* could also be used for other experiments. The *SimCrashLayer* is appropriate in cases of injection of faults: during "crash periods", the upper layers are isolated from the distributed system, thus appearing as crashed. The *Multiplexer* is instead useful when we want to fairly compare different distributed protocols with the same sending and receiving interface, guaranteeing that they perceive identical network (or lower level) behavior.

---

[2] Japan Advanced Institute of Science and Technology.

With the experiment described here we obtained interesting results about the mechanisms of failure detection for WANs, in terms of quantitative Quality of Service. In [13] there is a complete description of the experiment and of the obtained results.

The structure of the experiment highlights the potentiality and the easiness of an analysis performed using Neko and NekoStat. In absence of the addition of the NekoStat extension, the same results are obtainable, but with more difficulty. Without NekoStat it is necessary to collect the traces of execution of the processes, extracting quantitative information about the metrics using an ad-hoc tool and finally we have to use another tool to extract statistical parameters of the obtained measurements.

As a more general comment, not specific to the analyzed case study but to the general Neko and NekoStat framework, we remark that the main limitation of the experimental evaluations made with NekoStat is the impossibility to exactly define the temporal behavior of Java applications, essentially caused by the run-time garbage collection mechanism of the Java Virtual Machine. NekoStat support classes are implemented in Java and they are executed in the JVM: the mean and the maximum time necessary to execute a single action can be very different (e.g.: insertion of a new Event in the local EventCollector). This can thus influence the *accuracy* of obtained results: the garbage collection mechanism of the JVM can disturb/perturb the execution of the distributed system.

We are now extending the analysis by evaluating the QoS of the same failure detectors on different network conditions; namely on other WAN connections, and on LAN and WLAN environments. Repeating this experiment on different hosts is simple: it is sufficient to install Neko, NekoStat, and the specific experiment classes on the hosts that we want to use for the experiment. After that, we have only to change the configuration file for the Neko application, defining which are the hosts composing the distributed system. Also in the case that we want to evaluate the QoS of the failure detectors on a simulated network, we can use the same structure, layers and supports.

5   Conclusions and Future Works

This paper has presented the NekoStat tool, which is an extension to the already existing Neko framework for the analysis of distributed systems/protocols. Neko, although powerful and easy to use, allows only collection of traces of execution, and does not include any support to manage the gathered events to perform quantitative evaluations, in parallel with the protocol execution.

While retaining the appealing features of Neko, namely the ability to perform evaluations adopting both the simulative and experimental approaches, NekoStat enriches Neko with mathematical supports to handle the numerical quantities, as well as with analysis supports, to collect relevant distributed events and to analyze them on-line. The added value provided by NekoStat is in the direction of a statistical dynamic evaluation of a system, both on simulated and real execution, thus obtaining in an effective and easy way quantitative assessments of dependability, performance and, more in general, QoS metrics. Quantitative estimates obtained through NekoStat are beneficial under several aspects:

- they allow knowledge of the performance of a system with respect to specific application requirements;
- they constitute a valid support to the design and refinement of the protocol, by identifying weaknesses and bottlenecks;
- they allow comparison of several solutions, in order to select the most appropriate one.

We are currently working on devising additional extensions, both to Neko and to NekoStat, in order to further improve the analysis of distributed systems. In particular, two directions are under investigation:

- to extend the framework to include, as Neko layers, portions of source code of distributed algorithms written in languages different from Java (e.g., C and C++). Removing the restriction to use only algorithm written in Java will allow the analysis of a much richer population of already existing distributed protocols written in languages other than Java, without the necessity of any translation. Apart from easing the analysis process, this feature is very attractive especially in those cases where the translation in Java is not straightforward or possible at all (e.g., because Java does not contain support for some low-level functions). In any case, avoiding the translation improves efficiency and is less error-prone;
- to design new statistical analysis tools, to increase the analysis capacity of NekoStat.

Acknowledgments

Thanks to Peter Urbán and to Prof. André Schiper for Neko support and for useful comments during the development of the NekoStat extension.

References

1. Avizienis, A., Laprie, J., Randell, B., Landwehr, C.: Basic concepts and taxonomy of dependable and secure computing. IEEE Transactions on Dependable and Secure Computing 1 (2004)
2. Urbán, P.: Evaluating the performance of distributed agreement algorithms: tools, methodology and case studies. PhD thesis, Ecole Polytechnique Fédérale de Lausanne (2003)
3. Urbán, P., Défago, X., Schiper, A.: Neko: a single environment to simulate and prototype distributed algorithms. In: Proc. of the 15th Int'l Conf. on Information Networking (ICOIN-15), Beppu City, Japan (2001)
4. Urbán, P.: (Neko 0.9 website. http://lsrwww.epfl.ch/neko/)
5. Hoschek, W.: (Colt library website. http://dsd.lbl.gov/~hoschek/colt/)
6. Hoschek, W.: (Colt library api documentation. http://dsd.lbl.gov/~hoschek/colt/api)

7. Law, A.M., Kelton, W.D.: Simulation, Modeling and Analysis. McGraw-Hill (2000)
8. Falai, L.: Metodologie e strumenti per l'analisi quantitativa sperimentale e simulativa di algoritmi distribuiti. Tesi di laurea (in italian), Università degli Studi di Firenze (2004)
9. Mills, D.L.: Internet Time Synchronization: the Network Time Protocol. In: Zhonghua Yang and T. Anthony Marsland (Eds.), Global States and Time in Distributed Systems, IEEE Computer Society Press. (1994)
10. Verissimo, P., Rodrigues, L.: Distributed Systems for System Architects. Kluwer Academic Publishers, Norwell, MA, USA (2001)
11. Chandra, T.D., Toueg, S.: Unreliable failure detectors for reliable distributed systems. Journal of the ACM **43** (1996) 225–267
12. Chen, W., Toueg, S., Aguilera, M.K.: On the quality of service of failure detectors. IEEE Trans. Comput. **51** (2002) 13–32
13. Falai, L., Bondavalli, A.: Experimental evalutation of the QoS of failure detectors on Wide Area Network. In: Proceedings of the International Conference on Dependable Systems and Networks (DSN 2005), Yokohama (2005)

Airborne Software Concerns in Civil Aviation Certification

Benedito Sakugawa[1], Edson Cury[1], and Edgar Toshiro Yano[2]

[1] Industrial Fostering and Coordination Institute –IFI/CTA,
Praça Marechal Eduardo Gomes, 50,
12228-900 Vila das Acácias, São José dos Campos, SP, Brazil
{benedito.sakugawa, Edson.cury}@ifi.cta.br
http://www.ifi.cta.br
[2] Technological Institute of Aeronautic – ITA/CTA,
Praça Marechal Eduardo Gomes, 50
12228-900 Vila das Acácias, São José dos Campos, SP, Brazil
yano@comp.ita.br, http://www.ita.br

Abstract. In the civil aviation certification the software has an unlike treatment due to its peculiarities and also for being a relatively new item. There is no specific software certification requirement in the FAR[1] – FAR 33.28 is the only section that mentions the word *software*. The FAA[2] recognizes the considerations presented in RTCA/DO-178B[3] as an acceptable means for approval of software used in airborne systems for civil aviation. The CTA/IFI/CAvC[4], responsible for the type certification in Brazil, has been applying RTCA/DO-178B since it was issued. The purpose of this paper is to present the experience of CTA in applying DO-178B, focusing on those technical issues that were source of controversy among certification authorities and industries. This paper is relevant at present time as RTCA and EUROCAE[5] have recently organized a *Special Committee*[6] intending to issue DO-178C by the end of 2008.

1 Introduction

The civil aviation industry has a high level of safety due to careful investigation and analysis of accidents and immediate feedback of experience to design and operation. The conservatism approach, i.e., the use of well-matured technology, has contributed to keep the safety level. The appearance of embedded programmable electronic has threatened this paradigm - for instance, although its impact on safety is not well

---

[1] The United States Federal Aviation Regulation.
[2] The United States Federal Aviation Administration.
[3] The Radio Technical Commission for Aeronautics document called "Software Considerations in Airborne Systems and Equipment Certification".
[4] Aerospace Technical Center, Industrial Fostering and Coordination Institute, Civil Aviation Certification Division.
[5] European Organization for Civil Aviation Equipment.
[6] RTCA Special Committee SC-205 and EUROCAE Working Group WG-71.

C.A. Maziero et al. (Eds.): LADC 2005, LNCS 3747, pp. 52–60, 2005.
© Springer-Verlag Berlin Heidelberg 2005

known, the software has increased its presence in airborne systems and performed more critical functions [12].

The Brazilian civil aviation industry, started on the early 1970's, has been following the world tendency. Its products can be divided in four generations: the first used mainly analogical technology; the second had a small number of software items performing secondary functions; the third presented a great number of airborne electronics and almost 50 software items; the fourth and present generation has nearly 100 airborne software items, some performing critical functions. [11]

The certification authorities have expressed their concerns on establishing clear criteria for the use of software on critical airborne systems and equipments. The FAA recognized the considerations presented in RTCA/DO-178B [1], as an acceptable means for approval of software used in airborne systems for civil aviation. The CTA responsible for the type certification in Brazil has been applying DO-178B[7] since it was issued. The objective of this paper is to present the CTA experience in applying DO-178B focusing on those technical issues that were not either clearly addressed or in the scope of the guide and, therefore, have demanded extra effort from the certification authorities as well as the industries. First it shows the relationship between software and certification regulation, what it means by complying with regulation, the existing guidance documents for helping compliance, and the role of software in this compliance. Then it gives a brief description of DO-178B. After that, a selected list of technical issues is presented followed by discussions related to the list, and finalized with the conclusion and some references. This paper does not intend to detail the certification process, regulation, advisory materials and organizations involved, which would demand a specific work. Information on those topics can be found in [9] and [11].

2   Relationship Between Software and Certification Regulation

Certification regulation requires that the consequences of all failures should be analyzed. A catastrophic consequence should virtually never occur in the fleet life of an aircraft type, while less hazardous one is permitted to occur more often. The probability of a failure to occur should be inversely proportional to the severity of its consequence [13]. For certification, the assurance of an acceptable level of risk should consider:

1. *Number of failures*: No single failure can lead to a catastrophic consequence, independent on how unlikely is the failure occurrence;
2. *Probability*: For each hazard category there is either a quantitative or a qualitative measurement; and
3. *Design Evaluation*: Evaluation to confirm the absence of design errors.

For software only the design evaluation is required, which is the scope of DO-178B. Hardware usually requires all three considerations.

---

[7]   The European standard issued by EUROCAE is ED-12B which is exactly same as DO-178B, as RTCA and EUROCAE have worked together and reached a consensus.

3 Certification Guidance Documents

Certification regulation specifies levels of safety that are required. In order to verify if the safety requirements are met, a systematic system safety assessment must be performed. With this intent, various techniques have been developed and can be found in ARP4761 [4].

There is an additional safety concern for those called highly-integrated or complex systems: the existence of development errors (requirements determination and design errors). Guidance has been developed with the basic idea of ensuring that safety is adequately addressed through development process, and can be found in ARP4754 [3]. For complex hardware, refer to RTCA/DO-254 [2]. The figure-1 shows the relationship among these guidance documents and DO-178B.

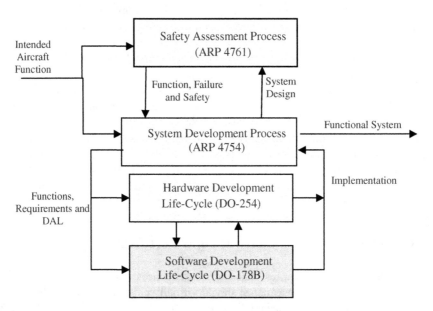

Fig. 1. Certification Guidance Documents Covering System, Safety, Software, and Hardware Processes [3]

An important output from the System Development Process is the Development Assurance Level (DAL) for system, software, and hardware. For ARP4754, development assurance is "all of those planned and systematic *activities* used to substantiate, at an adequate level of confidence, that development errors have been identified and corrected such that the system satisfies the applicable certification basis"[3]. The effort and detail needed in performing such *activities* depend on the DAL assigned to the system and its items (software and hardware), which is based on most severe failure condition classification associated with aircraft-level functions fully or partially implemented in them. For DO-178B, DAL is equivalent to software level. The table-1 presents a list of failure condition classification, the acceptable probability interval for their occurrences, and the corresponding software level required. For a detailed definition of failure condition classification, refer to AC 25-1309-1A [6].

Table 1. Failure Conditions and Respective Software Levels

Failure Condition	Probability[8]	Software Level
Catastrophic	$< 10E^{-9}$	A
Hazardous	$< 10E^{-7}$	B
Major	$< 10E^{-5}$	C
Minor	$< 10E^{-3}$	D
No Effect	Any	E

The *Software level* column on table-1 could also be applied to partitions. As explained in the DO-178B, partitioning is a technique for providing isolation between functionally independent software components to contain and/or isolate faults and potentially to reduce the effort of the software verification process.

4   The RTCA/DO-178B

The *RTCA/DO-178B "Software Considerations in Airborne Systems and Equipment Certification"* provides recommendations for the production of software. It presents guidance for determining, in a consistent manner and with an acceptable level of confidence that the software aspects comply with certification regulation. It presents the processes of planning, development (requirements, design, coding, integration), and integral processes (verification, configuration management, quality assurance, certification). The software life cycle, transition criteria, life cycle data generated, and additional considerations (software reuse, tool qualification, alternative methods) are also described. A list of 66 objectives is described, and if the developer can demonstrate satisfaction of these objectives, the software will be approved.

Table 2. Number of Objectives for Each Process

Process →	Plann.	Develop	Verific.	Config. Ctrl	Quality Assur.	Certific. Liaison	Total
Numb. of Objectiv.	7	7	40	6	3	3	66

The greatest effort is spent on verification, a technical assessment of both the software development processes and the software verification process, and comprises activities like reviews, analyses and test. The table-3 presents the number of objectives for each software level, listing whether the objectives should be satisfied with independence or not. In this context, independence means that the verification

---

[8] It is not acceptable to assign probabilistic numbers to software levels.

activity should be performed by a person other than the developer of the item being verified.

Table 3. Number of objectives for each software level. Note: With = with independenceWout = without independence.

Software Level	Number of Objectives		
	With	Wout	Total
A	25	41	66
B	14	51	65
C	2	55	57
D	2	26	28

The higher is the software level, the more rigorous is the guide, i.e. more objectives to satisfy. The DO-178B does not provide any guidance for software classified as level E, as there is no safety impact.

5 Relevant Technical Issues

Following is a summarized list of technical issues that were selected for being source of discussions and debates among certification authorities and industries. They are called technical issues because those managerial and organizational issues related to certification process were not included.

- **Issue#1:** The DO-178B is sometimes misunderstood as a software development standard.
- **Issue#2:** There are no clear criteria for acceptance of software reuse, new techniques like object-oriented technology, and automated tools.
- **Issue#3:** In many cases the software verification process was confined to the developer's environment that usually was not the aircraft manufacturer. Therefore, there was a risk of some software requirements being verified without enough evidence that they fully complied with system or even aircraft requirements.
- **Issue#4:** Some problems whose cause was attributed to software were actually consequence of deficiency in system requirements specification related to software.
- **Issue#5:** Some software items belong to a highly-integrated or complex system, and to show the software compliance individually to its own requirements may not be enough evidence that the system functions correctly, which implies several software and hardware items interacting adequately.
- **Issue#6:** The scope of software configuration control is limited to the set of software belonging to a single system, but the high integration among systems cause dependence among software from different systems, demanding a necessity for software control at aircraft level.
- **Issue#7:** Due to timing constraints, it has been a common practice to use for certification tests, versions of software that were still under acceptance process.

- **Issue#8:** Currently, there is no safety assessment process at the software level.
- **Issue#9:** Current certification guidelines allow software to be a single point of failure.

6 Discussion

This section presents a brief discussion on those issues introduced in section 5. It does not intend to be conclusive, but a starting point for further discussions.

On Issue#1: There are cases of applicants (or developers) that attempt to get acceptance of software modifications by claiming for credits from development processes previously audited and accepted by certification authorities. However, certification authorities do not certify processes, and for a clear reason: DO-178B is not a development standard that can create a software development environment, but an assurance standard. Assurance standards specify the characteristics that must be present in a development, but do not specify how to create those characteristics. The *how* belongs to development standards, which provide guidelines to ensure an orderly and repeatable development process. The idea is for the developers to choose a development standard for creating their software development environment, and then use the DO-178B to ensure that all needed visibility and characteristics have been captured by the development.

On Issue#2: The DO-178B was issued in 1992 and reflects the necessities of that time, when software used to be developed for specific application using well-known techniques and tools. That means, software reuse, object-oriented approach, and automated tools did not use to be a concern, as they were not widely utilized. The scenario has changed, and to fill up the gap the certification authorities (mainly FAA) has generated guidance material like Service History Handbook [8], Handbook for Object-Oriented Technology in Aviation [7], and Software Approval Guidelines [5], the latter addressing field-loadable software, tool qualification, use of Commercial Off-The- Shelf (COTS) software, and others.

Remark: The issue may be addressed by DO-178C.

On Issue#3: The DO-178B is organized with the assumption that the applicant is the software developer and does not consider the situation where they are separate entities, no rare located in distinct countries. In such case, an additional concern should exist to ensure the continuity in configuration management and quality assurance, because the test may migrate from the developer to the aircraft integrator (applicant) in order to fully verify those software requirements that demand the aircraft (or integrated systems bench) as a more adequate test environment.

Remark: The issue was addressed in the position paper CAST#11[9] [10].

On Issue#4 and Issue#5: The DO-178B is a development assurance standard, which assures a proper implementation of what was required by the system, but does not

---

[9] A paper written by the Certification Authorities Software Team (CAST). CAST comprises civil aviation software specialists from the United States, Europe, Canada and Brazil.

assure that the system requirements are correct. It does not tell how to design, verify and validate a system, which otherwise is the purpose of ARP4754 and ARP4761. They do provide guidance in the system architectural design against safety problems. ARP4754 basically adopt DO-178B approach and applied it for highly-integrated or complex systems. It comprises 11 chapters that deal with system development process, certification process and coordination, requirement determination and assignment of DAL, safety assessment process, validation of requirements, implementation verification, configuration management, process assurance, and considerations for modified aircraft.

On Issue#6: The scope of DO-178B does not cover the complete software life-cycle. Phases like installation, maintenance, and operation are not discussed. For example, the software in some cases is not considered an aircraft configuration item, but an internal "component" of an item, without individual identification (part number), making it very difficult to have a software configuration control at aircraft level that would contribute to more reliable installation, maintenance and operation.

Remark: The issue was partially addressed in Software Approval Guidelines [5].

On Issue#7: Three concerns can be identified: 1- a necessity for improvement of software configuration control at aircraft level, and in this particular case, during the certification test campaign, 2- a necessity of clear criteria for accepting use of versions still under evolution, and 3- a necessity of thorough analysis to verify whether new software versions would not invalidate past certification tests. Although at CTA all these concerns had been raised and coordinated by the software group, the execution demands more of system specialists than software, as the analysis is more functional and not of development assurance.

Remark: The issue is clearly out of the scope of DO-178B and will probably remain that way.

On Issue#8: The ARP4761 [4] describes a technique called Functional Hazard Assessment (FHA). FHA is performed at the functional level, usually when important information on software (e.g. architecture and design) is still unknown. Consequently, it does not address software safety requirements and assurance level. For software that performs safety-critical functions it is essential to analyze the software requirements, architecture and design, to ensure completeness and correctness of hazards identified by FHA, which means, a necessity for reassessing at software level the results of FHA. For example, identification of software failures which confirms the results of FHA or which could raise new causes for the hazards identified at the FHA level. In this way, software safety assessment would verify the results of previous system safety assessment steps.

Remark: The issue may be addressed by working group S-18[10], which is currently reviewing ARP4754 and ARP4761.

On Issue#9: Is it really possible to design software good enough to perform a function that has a potential to solely create a catastrophic event? The question is equally valid for hardware when design assurance is considered - only for random

---

[10] A working group coordinated by SAE-The Engineering Society for Advancing Mobility.

failures hardware is easily quantifiable, differently from software. For example, a structural component may have a single design error that could result in multiple structural problems throughout the aircraft, and the present regulations accept it. But that component has specific physical limitations, which allow it to fail in limited ways, making it possible for designers to mitigate the consequences. For software, however, there are no similar limitations, and it can fail in unpredictable ways. Software complexity has introduced additional concerns over other technologies that could justify not accepting it as a source of single point failure for catastrophic event. But on the other hand, the statistic indicates that software design errors, compared to other technologies, is not a major source of single point failure in the aircraft.[14]. What to do?

Remark: The issue is still a source of debates and controversy.

7 Conclusion

Although DO-178B is the main guidance for acceptance of airborne software used in civil aviation, some software concerns having direct impact on safety are not either clearly addressed or in the scope of DO-178B. In that case, it is necessary to use additional guidance material, which is provided by certification authorities (mainly FAA), some organisations (e.g. RTCA, SAE, EUROCAE), and dedicated technical groups (e.g. CAST). Both certification authorities and industries have recognized the need for some guidance documents update. International working groups have already started, hoping to solve deficiencies, some of them expressed on the issue list presented herein.

Based on CTA experience on civil aviation type certification, the paper presented some selected software issues that were cause of controversy and debates among certification authorities and industries. For technical support, it provided a brief description of the relationship between software and certification regulation, the certification guidance documents, and the assurance standard (DO-178B) used for airborne software development.

References

1. RTCA/DO-178B: Software Considerations in Airborne Systems and Equipment Certification. Radio Technical Commission for Aeronautics - RTCA, Inc. (1992)
2. RTCA/DO-254: Design Assurance Guidance for Airborne Electronic Hardware. RTCA, Inc. (2000)
3. SAE/ARP 4754: Certification Considerations for Highly-Integrated or Complex Aircraft Systems. SAE-The Engineering Society for Advancing Mobility, ARP-Aerospace Recommended Practice (1996)
4. SAE/ARP 4761: Guidelines and Methods for Conducting the Safety Assessment Process on Civil Airborne Systems and Equipment. (1996)
5. Order 8110.49: Software Approval Guidelines. Federal Aviation Administration – FAA (2003)
6. Advisory Circular AC 25.1309-1A: System Design and Analysis. FAA (1988)
7. Handbook for Object-Oriented Technology in Aviation. FAA (2004)

8. Software Service History Handbook. FAA (2002)
9. Cury, E., Sakugawa, B.M., Teixeira, M.M.R.: Certificação de Software Embarcado na Aviação Civil - Experiência Brasileira. CTA/IFI/CAvC, 2ª Safety Workshop, EPUSP (2004)
10. Position papers of the Certification Authorities Software Team - CAST http://www.faa.gov/certification/aircraft/av-info/software/software.htm
11. Lemes, M.J.R., Domiciano, A.J.V., Altoé, F.O., Carbonari, A.J.: Certificação de Software Embarcado de Emprego Aeronáutico: Processo e Desafios. Embraer, 1ª Safety Workshop, EPUSP (2003)
12. Leveson, N.G.: Safeware - System Safety and Computers. University of Washington, Addison-Wesley (1995)
13. Lloyd, E., Tye, W.: Systematic Safety. Civil Aviation Authority London (1982)
14. NASDAC – The National Aviation Safety Data Analysis Center, website, http://www.nasdac.faa.gov

A Method for Modeling and Testing Exceptions in Component-Based Software Development

Patrick Henrique da S. Brito, Camila Ribeiro Rocha, Fernando Castor Filho,
Eliane Martins, and Cecília M. Fischer Rubira

Institute of Computing - State University of Campinas,
PO Box 6176 ZIP 13084-971, Campinas, SP, Brazil
Phone: +55 19 3788 5842 / Fax: +55 19 3788 5847
{patrick.silva, camila.rocha, fernando, eliane, cmrubira}@ic.unicamp.br

Abstract. The design, implementation and testing of the exceptional
activity of a software system are complex tasks that usually do not receive the necessary attention from existing development methodologies.
This work presents a systematic way to deal with exception handling,
from the requirement specification phase to the implementation and testing phases, in component-based software development. Testing activities
are performed since the early stages of development, promoting an increase in the quality of the produced system. Our solution refines the
Methodology for the Definition of Exception Behavior, MDCE, in the
architectural design, implementation, and testing phases. Moreover, the
proposed method was adapted to the UML Components process.

1 Introduction

In order to reduce the cost and time-to-market of large software systems,
component-based software development (CBD) is rapidly gaining wide acceptance. Its adoption is motivated mainly by the extensibility and reuse of code at
a high level of granularity promoted by the use of this technology [35,9]. Due to
its popularity and potential benefits, CBD is currently being used in the development of computing systems with strict dependability requirements, such as,
mass transportation controllers and automotive devices.

The activity of a program is normal when it behaves according to its functional specification. When the program presents deviations from its functional
specification, its activity is said to be abnormal or exceptional, since it is expected that these deviations occur only rarely. Exception handling [18] is a well-
known technique for structuring the exceptional activity of software systems.
It is implemented by many mainstream programming languages, such as C++,
Java, and C#. In spite of the popularity of exception handling, the design, implementation and testing of the exceptional activity of a software system are
complex tasks that do not receive the necessary attention from existing development methodologies [12,30]. As a consequence, developers do not use exception
handling mechanisms appropriately, do not focus on designing the exceptional

C.A. Maziero et al. (Eds.): LADC 2005, LNCS 3747, pp. 61–79, 2005.

activity of applications, and do not cover appropriately this behavior during the test phase, therefore compromising overall system reliability [27].

The use of CBD in the development of critical systems highlights the importance of considering the exceptional activity and the overall system quality through validation techniques. This work presents a method, called MDCE+, for the definition and testing of the exceptional activity of component-based software systems. Our solution refines the Methodology for the Definition of Exception Behavior (MDCE) [30], which is an extension of the Catalysis [14] process for CBD. MDCE presents guidelines for the specification of exceptional activity of a system since the early phases of development and, begin focused on the requirements definition and analysis phases. MDCE+ refines MDCE mainly on architectural design, implementation, and testing phases, since the latter did not cover these phases in depth.

Testing activities, most of which can be automated, are distributed amongst all development phases in order to improve the reliability of the produced system. There are two main activities: testability improvement, embedding built-in testing capabilities into the component under test, and test cases generation, following the model-based approach [6,4]. All the testing artifacts can be reused each time the component is tested: during its development or each time it is reused.

Development and testing were integrated in previous component-based methodologies, as proposed by Atkinson et al. in KobrA [2,3] and by Farias [15]. KobrA uses inspections and testing as quality assurance policies. They propose inspections for the artifacts produced, system testing, and component testing based on built-in contract-based testing. Farias' approach deals only with system testing, presenting guidelines for automatic test case creation and execution. None of the two methodologies presents specifics activities for defining or testing the exceptional activity of the system.

To the best of our knowledge, the only other work that proposes a methodology for defining the exceptional activity of a system since the early phases of development focuses on object-oriented systems [25,12]. Furthermore, Sinha and Harrold [33] propose an approach for testing the exceptional activity of a system in a white box way. This work only covers unit tests and requires the source code of the tested components to be available.

In addition to refining the MDCE, we have adapted it to the UML Components [9] CBD process, mainly due to the fact that it has a simple structure and is easy to learn and apply in practice. These features make it more accessible to the corporate market, especially when compared to other CBD processes, such as Catalysis.

The rest of the paper is organized as follows. Section 2 briefly presents the concept of idealized fault-tolerant component and the UML Components process. Section 3 presents the MDCE+ method, its main characteristics, and the adaptations made to the UML Components process. Section 4 describes the development of a real financial application using the method and presents some examples of the artifacts produced in each phase. Section 5 presents a preliminary evaluation of the method, based on the results of the case study. Section 6 presents some concluding remarks and directions for future works.

2   Background

2.1   A Dependable Software Architecture Based on Exception Handling

Following the terminology adopted by Lee and Anderson [1], a system consists of a set of components that interact under the control of a design. Software components receive service requests and produce responses, which can be separated into two distinct categories: *normal*, which correspond to those situations where the component has provided its normal service satisfactorily; and *exceptional*, usually signaled when an error is detected, and the component cannot provide the requested service. Exceptional responses are usually called exceptions [18].

Exceptions can be classified into two different categories: *internal*, raised by a component in order to invoke its own error recovery measures, and, if this exception is handled successfully, the component can return to provide its normal service; and *external*, signaled if a component determines that, for some reason, it cannot provide its specified service. External exceptions can be partitioned into *interface exceptions*, which are due to an invalid service request, and *failure exceptions*, which are due to a failure in the processing of a valid request. In this sense, exceptions and exception handling provide a suitable framework for structuring the fault tolerance activities incorporated in a system.

Figure 1 presents the idealized fault-tolerant component [1] (IFTC), a structuring concept for building fault-tolerant systems by means of exception handling techniques. An IFTC promotes separation of concerns between the normal activity of a system and its exceptional activity, where measures for fault tolerance are implemented. An IFTC produces three types of responses: (i) normal responses; (ii) interface exceptions; and (iii) failure exceptions.

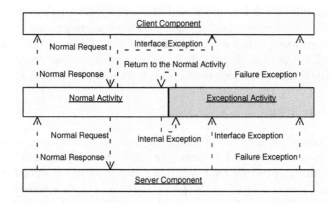

Fig. 1. Idealized fault-tolerant component (IFTC) [30]

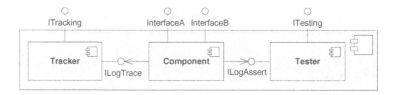

Fig. 2. Testable Component Architecture

IFTCs may be organized into layers, so that components may handle exceptions raised by components located in other layers. In this approach, the system software architecture is partitioned in layers that comprise different levels of abstraction. Ideally, each layer is responsible for handling only exceptions raised by the layer immediately below it.

2.2   Testable Component Architecture

Building components with good testability can simplify test tasks and reduce test costs. Our testable component architecture [29], illustrated in Figure 2, augments the component with test facilities known as built-in tests (BIT) [5], which are accessed by the user through a standard interface.

The monitoring interface implements services to embed monitoring capabilities in the component under test, to monitor methods and attributes/properties. Monitoring facilities can be useful not only for testing, but also for debugging purposes, or when a component is tested in an environment for which there is no possibility to have drivers or stubs interacting directly with the component.

The test interface implements services to embed built-in test capabilities in the component under test. These capabilities concern components contract verification at runtime, following the Design by Contract approach [26], which are used as test oracle. Besides, this interface is also responsable for retrieving behavior models of the component, which are used during test case generation.

There are also internal interfaces for logging, which are responsible for reporting the monitoring information and the assertion violations in a log file. Both services are implemented using aspect-oriented programming [21], and were structured as an aspects library inside Tracker and Tester components. With this technique, it was possible to embed the code even without the component's source code, making reutilization possible even if the source code is not available.

2.3   The UML Components Process

UML Components is a software development process which focuses on the construction of component-based systems. In order to simplify development, it adopts a specific architecture that highlights two layers: a *system layer*, which comprises components related to the particularities of the software system, and a *business layer*, comprising components that can be reused across different applications. Development is divided into six phases, described as follows.

The first phase of UML Components is *requirements specification* , when the developer specifies the functional requirements of the system as a set of *use cases* [31]. Moreover, the *business concept model* is specified, which represents the basic entities of the conceptual domain of the system.

The following phase is *component specification*, the most important one, which is divided into 3 subphases: (i) *component identification*, when the developer identifies system layer interfaces and their respective operations based on use case descriptions, and also business interfaces, defined from the core entities (or *core types*) of the business concept model; (ii) *component interaction*, when the operations of the business layer, which are required by the system layer, are defined; (iii) *final specification phase*, when the interfaces are refactored and operation execution contracts are formalized through pre and postconditions [26].

Next is *provisioning*, when components are either built from scratch or acquired from third-parties. These components are later combined in order to realize the architectural configuration of the system during the *assembly* phase. UML Components does not provide any guidelines on how the *testing* and *deployment* phases should be executed.

3   The MDCE+ Method

The MDCE+ method systematizes the identification, design, implementation, and testing of the exceptional activity in the software development phases.

The activities of the MDCE+ method were distributed among the phases of UML Components, as illustrated in Figure 3. The left column shows the main UML Components extension points concerning exceptional activity definition, and the right column shows the testing activities included.

The exceptional activity definition starts already in the requirements specification phase, where exceptions are identified. During the specification of the exceptional activity, exceptions are classified according to the IFTC model. Moreover, according to the way they are detected and handled, they can be classified as: *internal*, when they are raised by the component in order to invoke its own exceptional activity, or *external*, when they are signaled by the component if it determines that, for some reason, it cannot provide its specified service. An external exception can be classified as *architectural* when it is signaled by an architectural component.

Architectural exceptions cross the boundary between two architectural components, which means that the architectural exceptions that flow between two components are part of the interaction protocol to which these two components adhere. Because MDCE+ handles architectural exceptions, the importance of architectural connectors [32], which realize the interactions between architectural components, is highlighted. Besides handling this kind of exception, these connectors are also responsible for detecting context-dependent exceptions.

MDCE+ testing activities, so far, only concern component testing level, which follows the model-based responsability testing approach [6]. In this level, we are concerned on checking if the component behaves like it was designed,

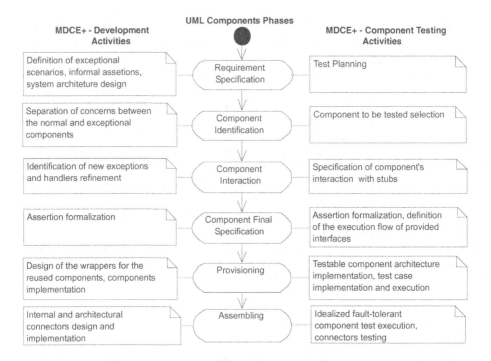

Fig. 3. MDCE+ interference in UML Components's phases

including its normal and exceptional parts. In the future, robustness testing techniques such as [23,22,13], will be incorporated in integration and system testing levels.

The component testing is performed in a black-box way, allowing test case reutilization even without component source code. Testing activities also starts during requirements specification, with test planning, and are distributed throughout all development phases, consisting of test case generation and component testability improvement.

Test cases are generated from component behavior models, produced during component specification phase. Although in these case study the test cases were developed manually, the steps described in the following sections can be automatized by a future tool.

Testability [5] concerns how easy is to test a system, contributing to test eficiency. In this case, component testability is improved with the inclusion of tracking and contract verification mechanisms in the component under test, as presented in Section 2.2. Tracking mechanisms can ease faults localization and decrease maintenance costs. Contract verification during runtime can act as a test oracle, which means the expect result of the test case. As described later in the text, these mechanisms can also be automatically generated from specifications by future tools.

The following subsections describe the activities included in each of the UML Components phases.

Requirements Specification and Architectural Design. The main objectives of this phase are to discover and specify the functional requirements and the attributes of quality (non-functional requirements) of the system. The developer is responsible for perfoming the activities prescribed by UML Components (defining the business concept model and specifying use cases) and some new activities introduced by MDCE+. The latter comprise defining exceptional scenarios, which describe error situations and how to handle them, and extend the specifications of use cases with invariants, pre and postconditions, following the Design by Contract methodology.

The main purpose of the new activities introduced by MDCE+ is to discover exceptions early in the development process. Contracts are important because exceptions can be anticipated by analyzing the possible violations of these contracts. Exceptional scenarios show what exception handlers should do. The business concept model is also used for defining exceptions since, based on this model, specialists on the application domain can identify the most critical entities. These critical entities will probably be realized as components that implement some form of redundancy, either architectural exception handlers or some mechanism for design diversity [1]. Identification of critical entities also helps test planning issues, such as test schedule, resources, and people.

Another important activity introduced by MDCE+ is the UML Components architecture customization, according to system requirements. During this customization, the architectural layers internals can also be detailed and may possess different architectural styles.

Component Identification. According to UML Components, this phase starts with the identification of the provided interfaces of the system layer and the components of the business layer. For system layer interfaces, operations are discovered examining the steps of the use cases. The last activity is grouping these interfaces as provided component interfaces, observing the component cohesion.

MDCE+ extends the Component Identification phase with two activities: (i) definition of exceptions and handlers; and (ii) selection of candidate components for unit testing.

In the first activity, an exceptional class and an exceptional interface are created for each exception identified in the requirements specification phase. *Exceptional classes* encapsulate contextual information regarding the errors that trigger the exceptions; and *Exceptional Interfaces* define methods that implement different exception handling strategies for the different contexts in which the exception may be caught, and are grouped into exceptional components.

In the second activity, the developer selects components to be tested as a black-box. Components marked as critical have a higher priority because their dependability is crucial to the system. The same applies for reusable components, because they will be employed many times. For both cases, it is recommended that test suites be devised, so that regression testing can be performed automatically. The components not selected during this phase will be tested during the integration and system testing phases.

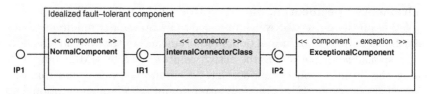

Fig. 4. Realization of the IFTC

Component Interaction. In this phase, according to UML Components, business operations are identified using UML colaboration diagrams. In the MDCE+, colaboration diagrams were replaced by UML activity diagrams [19], because they can be also used for test case generation (stub creation), as it illustrates the sequence which the required operations are called. Test generation will be exemplified in Section 4.

New exceptions can be discovered as well, by evaluating unforeseen conditions resulting from the steps of the scenarios. Exception propagation between architectural components can also be analysed. These exceptions can be handled in the architectural connectors or further propagated. Developers can analyze exception propagation in software architectures by hand, using scenarios, or automatically, using tools such as the Aereal [16] framework.

Finally, architectural components are structured as IFTCs. In this step, each normal component is associated to exceptional components, which provide handlers for exceptions that may be signaled by the operations in the provided interfaces of the normal component. A specific internal connector realizes this association, as shown in Figure 4. As for testing, the list of components to be tested is reviewed based on the identification of IFTCs. It is also necessary to decide whether the normal and exceptional components will be tested separately or as a single entity.

Component Final Specification. In this phase, as MDCE+ activity, the normal and exceptional interfaces can be refactored, in order to reduce the number of interfaces, without sacrificing cohesion. Possible violations of contracts between the provided and required components interfaces are identified, and are normally handled in the architectural connectors through conversion of exception types.

Afterwards, test models are specified. The first test model is the specification of the execution flow of provided interfaces, using UML activity diagrams, as proposed in [8]. This model will be used for test case generation and illustrates the sequential dependencies between the methods in the provided interfaces. These sequential dependencies define the component behavior, both normal and exceptional.

The other test model produced in this phase is the formal contract of the component, based on the informal assertions specified during the requirements specification phase. Contracts are formalized in UML OCL (Object Constraint Language) [20], in order to automatically generate contract verifications at runtime [7].

Provisioning. In this phase, normal and exceptional components (not the IFTC) are made available separately. According to UML Components, there are three ways to obtain the components that will be used to implement a system: (i) reuse of existing components; (ii) acquisition of Commercial Off-The-Shelf Components (COTS Components); (iii) implementation of new components.

When a component is reused or acquired, MDCE+ foresees the creation of adapters, which may be implemented either via wrappers or inside the architectural connectors themselves (in the assembly phase). For implementing new components, MDCE+ proposes the use of a system implementation model that explicitly materializes architectural components at the implementation level, namely COSMOS [11]. Its details are explained in Section 4.

The exceptional classes should also be organized according to the hierarchy shown in Figure 5 [17]. This hierarchy defines exception types aiming to relate internal and external exceptions consistently. We can map the classification given by the IFTC to the types in the hierarchy of Figure 5 as follows. *Interface exceptions* inherit from RejectedRequestException. *Failure exceptions* inherit from DeclaredException if they refer to errors that are part of the specification of the system (expected errors). Failure exceptions inherit from UndeclaredException if they refer to error conditions that are not addressed in the specification of the system (unexpected errors). UndeclaredException has two subtypes that specify the state in which a component was left after the exception was signaled. A failure exception inherits from RecoveredFailureException if it is known that the component was left in a consistent state after the exception is raised, for instance, because it implements some backward error recovery mechanism. Conversely, a failure exception inherits from UnrecoveredFailureException if it is not known whether the system is consistent after the exception is raised. Internal exceptions can be of any type, as long as they are converted to some exception in the exception type hierarchy when they reach the boundaries of the component.

After the implementation of each component, the components behavior models are revised and testability improvement mechanisms and test cases are produced. Testability improvement mechanisms are produced based on interfaces specification (tracking mechanisms) and OCL contracts, produced in the com-

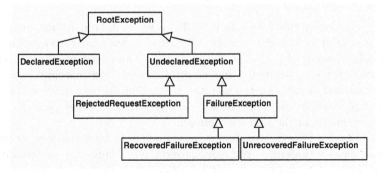

Fig. 5. Hierarchy of Suggested Exceptions [17]

ponent final specification phase. These mechanisms can be implemented by using the *testable component architecture* presented in Section 2.2, which uses aspect-oriented programming [21] techniques to introduce tracking and contract verification mechanisms in the intermediate code of the component under test.

Test cases are derived from the activity diagrams built during the component interaction (interaction diagram) and final specification (execution flow diagram) phases. Both diagrams are associated recursivelly: each operation in the execution flow is detailed by the corresponding interaction diagram, that models the method execution flow concerning interactions with required interfaces. The execution flows diagram derive test drivers, which execute test cases, and the interaction diagram derive stub synchronization commands.

Stubs replace required components, simulating their behavior in a controlled way, and making it possible to observe component behavior under test in normal and exceptional situations related to interactions with required interfaces. Test cases are derived from diagrams paths [24], starting in the execution flow initial node, passing throw interaction paths, and ending in each corresponding final nodes, covering paths in both diagrams. Tests are executed in this phase for components that have already been produced.

Assembly. Due to the focus on the connections between components, the assembly phase basically consists of the architecture configuration realization, which means connectors design and main program implementation. Two kinds of conectors are produced: *internal*, which integrates the normal and exceptional parts of IFTCs; and *architectural*, which connect two or more IFTCs and where architectural exception handlers are implemented. In order perform assembly, all the normal and exceptional components have to be already available.

As for testing, IFTCs are tested using test cases generated in the previous phase for IFTC's normal component. Connectors are tested similarly to components. Complex connectors are tested in isolation, with the production of the models and instrumentation mechanisms necessary for test generation and execution. Simpler connectors are tested only in the context of the whole system, during integration and system testing.

4   Case Study

The MDCE+ was applied in a case study in an industrial environment. The method was used for the development of part of a financial system with strict dependability requirements. This system registers and controls the delivery of check-books, account contracts and credit limits. It was specified using the MDCE+ method along with the adapted UML Components process. The case study was planned by the authors and executed by two other persons, one of them is a specialist in the business domain.

The main goal of the case study was to evaluate fault tolerance aspects such as exceptions quality (independence from programming language or development platform) and fault treatment quality (critical exceptions that behave transparently from the customer point of view). Beyond the fault tolerance-related criteria, other general aspects were also analyzed, such as maintainability and testability. The next subsection describes part of the artifacts that were created, demonstrating our method, and conclusions about the analyzed aspects.

4.1   Execution

Requirements specification and Architectural Design. In the requirements specification phase, the first activity was the development of the business concept model, where 22 entities were identified and four were considered critical: Account, AgencyControl, BankPartners, and FinancialTransaction. Six use cases were specified and, in this paper, for the sake of simplicity, we focus on the Cancel Contract use case and one of its exceptions as an example. A very detailed description of the case study and the obtained results is available elsewhere [10].

The Cancel Contract use case was specified with normal, alternative and exceptional scenarios. For each scenario, pre and postconditions were specified. For example, one of the preconditions of the main scenario of Cancel Contract was "Account agency must be registered". Based on violations of this assertion, the Not Registered Agency exception was derived.

Another activity performed in this phase is the design of the system's software architecture. This activity must obey some restrictions imposed by UML Components. The architecture should comprise at least one system architectural component, which is related to the particularities of the software system, and a business architectural component, formed by components that can be reused across different applications.

Besides the aforementioned restrictions, the architecture adopted for the software system, presented in Figure 6, reflects the infrastructure which exists in the company where the system was developed. The adopted architecture adhere to a

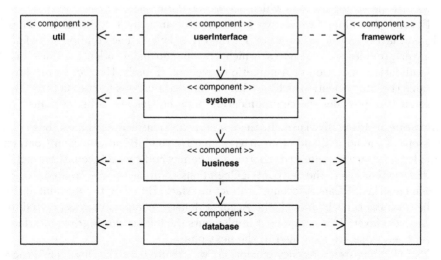

Fig. 6. System Architecture

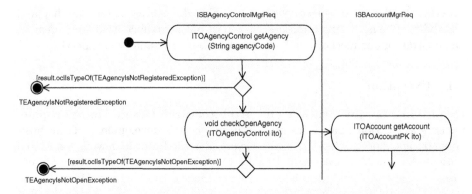

Fig. 7. Part of interaction diagram for cancelAccountContract method

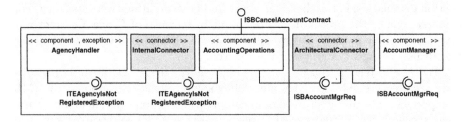

Fig. 8. Software architecture for the IdealizedAccountingOperations component

"relax posture" of the architectural layers [32] style. This architectural style has 4 traditional layers, which obey the communication constraints that are specified in the architectural layer style. The **database** layer contains the components which provide the operations which access the data bases.

Beyond these four layers, two layers had been defined additionally. These aditional layers can be accessed by all the other layers. The components of the util layer provides some services, which are domain independent, as an e-mail field validation operation or a numeric conversion. Finally, the *framework* layer contains the components which provide the needed business infrastructure. This infrastructure provides the communication between this and other systems.

Component Identification. In the component identification phase, based on the steps of the normal and alternative scenarios, the ISBCancelAccountContract provided interface was created, containing the method cancelAccountContract(). For the system layer, the AccountingOperations component was created, which implements the ISBCancelAccountContract interface. Based on the Account entity of the business concept model, the AccountManager component was created for the business layer. This component implements the ISBAccountMgrReq interface, whose operations are identified in the next phase.

The Not Registered Agency exception, was realized as the class TEAgency-IsNotRegisteredException, which keeps the agency code as context, and the ex-

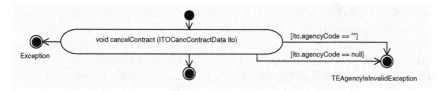

Fig. 9. Execution flow diagram for ISBCancelAccountContract interface

ceptional interface ITEAgencyIsNotRegisteredException. This interface was linked to the AgencyHandler component, which aggregates the exceptional interfaces related to agencies.

Component Interaction. In the component interaction phase, activity diagrams were created to specify the interactions in which the required interfaces participate. Part of the interaction diagram for the cancelAccountContract() method is shown in Figure 7. The swimlanes represent the required interfaces (ISBAgencyControlMgrReq, ISBAccountMgrReq), the actions represent the methods called, and the final nodes represent possible results of the cancelAccountContract() method. The values returned by required methods, including exceptions as TEAgencyIsNotRegisteredException, are registered as guard conditions in the actions following edges. Architectural exceptions concerning component reconfiguration were also identified in this phase.

As a result of component interaction definition, the connections between the components AccountingOperations, AccountManager and AgencyHandler were defined, as illustrated in Figure 8: AccountingOperations and AccountManager were connected via an architectural connector, and AccountingOperations and AgencyHandler were connected via an internal connector, in an IFTC form.

Component Final Specification. The activity in Component Final Specification phase was execution flow definition with activity diagrams, and contract formalization with OCL (Section 3).

Figure 9 illustrates ISBCancelAccountContract execution flow diagram. As in the interaction diagram, the final nodes define the expected results for the flow including exceptional ones as, e.g. TEAgencyIsInvalidException), related to and interface exception, and TEAgencyIsNotRegisteredException, an external exception also present in the interaction diagram (Figure 7). The method's parameters can be specified as guard conditions, including interface exceptions throwing. In this diagram.

In the OCL contract, invariant, preconditions and postconditions are formalized. The postconditions, as they can group verifications for various possible return values, are divided in condition/post expressions for each return value, similar as proposed in [9], meaning the return reason and return value.

Provisioning. After specification, the components AccountingOperations, AccountManager and AgencyHandler were implemented using COSMOS, which is detailed ahead. Figure 10 illustrates the AccountingOperations component internal structure. COSMOS defines three sub-models, which address different aspects of the CBD systems: (i) the *specification model (spec package)* specifies the interfaces provided and required by each component in the system, including the exceptional ones (the normal required interfaces were grouped into the ISBAccountingOperationsBusReq interface, and the exceptional ones into the IS-BAccountingOperationsExcReq interface); (ii) the *implementation model (impl package)* defines how the services provided by the component are implemented, including how the component is instantiated (component instantiation control classes are part of COSMOS); and (iii) the *connector model* (not illustrated, as it is an architectural model) specifies the connections between components using connectors, thus enabling two or more components to be connected in a configuration. Each of these models is implemented as a well-defined pattern which can be automatically translated to source code.

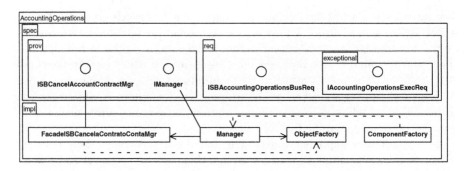

Fig. 10. AccountingOperations component internal structure

In this phase there was also testability improvement generation mechanisms, as described in [29], and test case generation from the UML activity diagrams. Figures 9 and 7 were glued, and the interaction diagram became a second level of the execution flow, illustrating the execution flow resulting from the invocation of cancelAccountContract() method. Five paths were extracted from the complete diagram by a depth-first search algorithm [28], producing five test cases.

One of them, e.g., simulates the TEAgencyIsNotRegisteredException throwing, and consists of four main steps: (i) the stub which simulates ISBAgencyControlMgrReq interface was prepared to throw the TEAgencyIsNotRegisteredException exception when the getAgency() operation were (DVIDA INGLS) called, was described in the portion of the path concerning the interaction diagram; (2) the component was created; (3) cancelAccountContract() method was invoked, as described in the execution flow diagram; (4) the expected value was checked as the TEAgencyIsNotRegisteredException exception throwing, as described in the path final node. Besides, there is also the contract verification, which checks exception class type and context.

Assembly. The last activity was connectors implementation, materializing the architecture showed in Figure 8. The internal and architectural connectors were implemented using COSMOS. The internal connectors links the AccountingOperations normal component and the AgencyHandler exceptional component, realizing the IFTC architecture, and the architectural connector links the IFTC AccountingOperations to AccountManager. As this connector implemented reconfiguration functions, it was also tested separately in a new MDCE+ iteraction.

4.2 Product Evaluation

Concerning the final product quality, approximately 20% of more exceptions were discovered with the systematic modeling of the exceptional activity, when compared with other equivalent systems in the company. This addition in the number of exceptions represents a bigger fault types refinement, that together with the more appropriate choice of the exceptions name (based on the assertions) and also with the use of their contextualized information, facilitate the maintenance activities related to the bugs identification and corrections.

Moreover, the proximity of the business logic and the independence of the specified exceptions related to the programming language, characterizing exceptions quality improvement, make a platform changing or a product line production easier. The exception contextualization information also contributed for fault treatment quality, as more detailed fault information could be presented to the customer.

Despite test case generation was performed manually, the testability improvement could be noticed as test case development was simply the translation of the diagrams paths to the JUnit framework. The contract verification was also effective as test oracle, capturing 90% of the defects observed. The tests coverage was over 80% both in normal and exceptional components, showing a high quality of the test cases, at least in what concern code coverage.

Because of the not critic aspects of some specified use cases, the case study was also important to evaluate how the method can be adequate to develop software systems, without critic dependability requirements. This adequacy was made in the component interaction phase, during the definition of exceptional super types. We have used these more generic exceptions to substitute the specific ones, satisfying project decisions.

5 Method Evaluation

Besides the method experience in only one case study, a preliminary analysis of the method could be performed, according to process quality criteria pointed by Sommerville [34]. The analysis was made by the authors together with the case study developers. The characteristics analyzed were:

Understandability. *Medium.* Besides the great number of activities, the clear separation between each activity facilitate the understanding.

Visibility. *High.* The results of all phases are clearly defined in documents, specially UML diagrams.

Supportability. *High.* All the models can be built using CASE tools, and both the COSMOS model source code and test cases code can be automatically generated from the models. Actually, a CASE tool for supporting the method, Bellatrix, is already been developed [36]. The Bellatrix environment will cover all method phases, supporting models production, code generation and test case execution.

Reliability. *Medium.* As process progress is all documented in the artifacts produced, and each artifact is used in the next phase, errors are generally captured before result in product errors. But some kinds of specification faults are not identified easily, for example, inconsistences between requirements and use cases specifications.

Robustness. *Low.* As there are not guidelines for project management, unexpected problems that make the development impracticable are not treated adequately.

Maintainability. *Medium.* As the activities are clearly separated through different phases, the inclusion of new activities is easy. The removal, although, is not that simple, as the each phase depend on the result of previous ones.

Rapidity. *Low.* As the main concern is the detailed specification of exceptional activity, the documents produced slow down the process.

An important contribution of the MDCE+ method is the adaptation of a practical software development CBD process (UML Components). New activities were added among the phases of this process, both concerning development and testing activities, as shown in Figure 3. These activities systematize the development and testing of reliable systems through specification, implementation and testing of the system exceptional activity.

6   Conclusions and Future Works

In this paper we presented MDCE+, a method for modeling and testing the exceptional activity of component-based software systems. MDCE+ improves the system dependability, providing a systematic way to modeling and testing the exceptions and its handlers, distributing its activities during all the software development cycle. This structured and rigorous way of detecting and handling exceptions in the context of faults occurrence is particularly relevant to those systems with high dependability requirements.

The main characteristic of MDCE+ method is the execution of development and test activities in parallel, following the model based testing approach. This parallelism reduces the overhead of its adoption, favoring the method application in real systems. Another important characteristic is the constant interaction between the developers and testers through specification artifacts, which are shared by the two teams. This interaction contributes for artifacts quality improvement. The synchronization is guaranteed by the joint update of these common artifacts.

Besides development and tests activities in parallel, MDCE+ emphasizes the architectural aspect of the system, contributing for the definition of exceptional flow of the system architecture. In relation to maintainability aspects, MDCE+ completely separates the specification of components normal and exceptional activity, and provides activities to compose these views in the Idealized Fault-Tolerant Component structure. This feature promotes better understandability, reliability and maintainability for the system, as these concerns do not get cluttered in a single component modeling.

The main limitation observed in the method was the necessity of update the specified artifacts. Lack of commitment of the development team can compromise the applicability of the generated test cases, implying in re-work and delay in the product deployment. To cope with this limitation, the MDCE+ method specifies joint update activities of the artifacts, which are run by both development and test teams.

Our most immediate future work is to build the process-oriented case tools that are oriented to MDCE+ method. These tools have to assist the construction of methods several artifacts, and test cases automatic generation. Moreover, currently only component testing activities have been specified, then another important future work is to complete the method to cover inspections, integration and system testing phases, where robustness testing techniques will be applied.

Finally, we intend to make a separation between the MDCE+ method and the activities of the UML Components process. This separation will makes possible the insertion of the same method in other software development processes.

Acknowledgements

We would like to thank the anonymous referees, who provided many interesting comments and suggestions. Patrick Brito is supported by the *lato sensu* postgraduate course in Software Engineering, IC/UNICAMP. Camila Ribeiro Rocha is supported by CAPES/Brazil. Fernando Castor Filho is supported by FAPESP/Brazil under grant 02/13996-2. Cecília M. Fischer Rubira is supported by CNPq/Brazil, grant number 351592/97-0.

References

1. Thomas Anderson and Peter A. Lee. *Fault Tolerance: Principles and Practice.* Prentice-Hall, 2 edition, 1990.
2. Colin Atkinson, Joachim Bayer, and Dirk Muthig. Component-based product line development: the kobra approach. In *Proceedings of the 1st conference on Software product lines : experience and research directions*, pages 289–309, Norwell, MA, USA, 2000. Kluwer Academic Publishers.
3. Colin Atkinson and Hans-Gerhard Gross. Built-in contract testing in model-driven, component-based development. In *Proceedings of the 1st International Working Conference on Component Deployment, Workshop on Component-based Software Development Processes*, 2002.

4. Antonia Bertolino, Eda Marchetti, and Henry Muccini. Introducing a reasonably complete and coherent approach for model-based testing. *Electr. Notes Theor. Comput. Sci.*, 116:85–97, 2005.

5. Robert V. Binder. Design for testability in object-oriented systems. *Communications of the ACM*, 37(9):87–101, 1994.

6. Robert V. Binder. *Testing object-oriented systems: models, patterns, and tools.* Addison-Wesley Longman Publishing Co., Inc., 1999.

7. L. Briand, W. Dzidek, and Y. Labiche. Using aspect-oriented programming to instrument ocl contracts in java. Technical Report SCE-04-03, Carleton University, 2004.

8. Lionel Briand and Yvan Labiche. A uml-based approach to system testing. *Software and Systems Modeling*, 1(1):10–42, September 2002.

9. John Chessman and John Daniels. *UML Components: A Simple Process for Specifying Component-Based Software.* Paperback, 1992.

10. Patrick Henrique da Silva Brito, Camila Ribeiro Rocha, Eliane Martins, and Cecília Mary Fischer Rubira. An integrated method for modeling and testing exceptions in component-based software development: A case study (in portuguese). Technical Report (to appear), Institute of Computing, 2005.

11. Moacir C. da Silva Jr., Paulo Asterio de C. Guerra, and Cecilia M. F. Rubira. A java component model for evolving software systems. In *Proc. of the ASE*, pages 327–330, 2003.

12. Rogério de Lemos and A. Romanovsky. Exception handling in a cooperative object-oriented approach. In *Proc. of the 2nd IEEE ISORC'99*, May 1999.

13. Regina Lúcia de Oliveira Moraes and Eliane Martins. Jaca - a software fault injection tool. In *DSN*, page 667. IEEE Computer Society, 2003.

14. Desmond D'Souza and Alam Cameron Wills. *Objects, Components, and Frameworks with UML The Catalysis Approach.* Addison-Wesley, 2nd edition, 1999.

15. C. Farias and P. Machado. A functional testing method for components verification (in portuguese). In *Proc. Brazilian Software Engineering Symposium (SBES)*, 2003.

16. Fernando Castor Filho, Patrick H. S. Brito, and Cecília Mary F. Rubira. A framework for analyzing exception flow in software architectures. In *Proceedings of the ICSE'2005 Workshop on Architecting Dependable Systems*, 2005.

17. Fernando Castor Filho, Paulo Asterio de C. Guerra, Vinicius A. Pagano, and Cecília Mary F. Rubira. A systematic approach for structuring exception handling in robust component-based software. *Journal of the Brazilian Computer Society - Special Issue on Dependable Computing*, 2005.

18. John B. Goodenough. Exceptional handling: Issues and a proposed notation. *CACM*, 18(12), 1975.

19. Object Management Group. *OMG Unified Modeling Language Specification Version 1.5*, 2003.

20. Object Management Group. *UML 2.0 OCL Specification*, 2003.

21. Gregor Kiczales, John Lamping, Anurag Menhdhekar, Chris Maeda, Cristina Lopes, Jean-Marc Loingtier, and John Irwin. Aspect-oriented programming. In Mehmet Akşit and Satoshi Matsuoka, editors, *Proc. European Conference on Object-Oriented Programming*, volume 1241, pages 220–242. Springer-Verlag, Berlin, Heidelberg, and New York, 1997.

22. P. Koopman et al. Ballista project : Cots software robustness testing. http://www.ece.cmu.edu/ koopman/ballista/, 2003.

23. N. P. Kropp, P. J. Koopman, and D. P. Siewiorek. Automated robustness testing of off-the-shelf software components. In *FTCS '98: Proceedings of the The Twenty-Eighth Annual International Symposium on Fault-Tolerant Computing*, page 230, Washington, DC, USA, 1998. IEEE Computer Society.

24. Eliane Martins, Cristina Maria Toyota, and Rosileny Lie Yanagawa. Constructing self-testable software components. In *Proc. DSN 2001*, 2001.

25. S. Messina and P. Pleinevaux. Enhancing cimosa with exception handling. In *Proceedings of the International Symposium on Robotics and Manufacturing ISRAM, World Automation Congress'96*, Montpellier, France, May 1996.

26. Bertrand Meyer. *Object-Oriented Software Construction*. Prentice-Hall, 1st edition, 1988.

27. Darrel Reimer and Harini Srinivasan. Analysing exception usage in large java applications. In *Proc. of ECOOP Workshop on Exception Handling in Object-Oriented Systems(EHOOS'2003)*, 2003.

28. Ronald L. Rivest and Charles E. Leiserson. *Introduction to Algorithms*. McGraw-Hill, Inc., New York, NY, USA, 1990.

29. Camila Ribeiro Rocha and Eliane Martins. A strategy to improve component testability without source code. In Sami Beydeda, Volker Gruhn, Johannes Mayer, Ralf Reussner, and Franz Schweiggert, editors, *SOQUA/TECOS*, volume 58 of *LNI*, pages 47–62. GI, 2004.

30. C. Rubira, R. de Lemos, G. Ferreira, and F. Castor Filho. Exception handling in the development of dependable component-based systems. In *Software Practice and Experience*. John Wiley and Sons, 2005.

31. Geri Schneider and Jason P. Winters. *Applying Use Cases: A Practical Guide*. Addison-Wesley, 1st edition, 1998.

32. Mary Shaw and David Garlan. *Software Architecture: Perspectives on an Emerging Discipline*. Prentice Hall, 1st edition, 1996.

33. S. Sinha and M. J. Harrold. Analysis and testing of programs with exception handling constructs. *IEEE Transactions on Software Engineering*, 26(9):849–871, 2000.

34. Ian Sommerville. *Software Engineering*. Addison-Wesley, 6 edition, 1995.

35. Clemens Szyperski. Component software and the way ahead. In Gary T. Leavens and Murali Sitaraman, editors, *Foundations of Component-Based Systems*, chapter 1, pages 1–20. Cambridge University Press, 2000.

36. Rodrigo Teruo Tomita, Fernando Castor Filho, and Cecilía Mary Fischer Rubira. Bellatrix: An environment to provide architectural support for component-based software development (in portuguese). In *IV Workshop of Component-Based Development (WDBC'2004)*, September 2004.

Verifying Fault-Tolerant Distributed Systems Using Object-Based Graph Grammars*

Fernando L. Dotti[1], Odorico M. Mendizabal[1], and Osmar M. dos Santos[2],**

[1] Faculdade de Informática,
Pontifcia Universidade Catlica do Rio Grande do Sul,
Avenida Ipiranga, 90619-900, Porto Alegre - Brazil
{fldotti, omendizabal}@inf.pucrs.br
[2] Real-time Systems Research Group, Computer Science, University of York,
Heslington, YO10-5DD, York - United Kingdom
osantos@cs.york.ac.uk

Abstract. Assuring the correctness of fault-tolerant distributed systems can be an overwhelming task. Besides dealing with complex problems of distributed systems, it is also necessary to design the system in such a way that a well-defined failure behaviour, or the masking of failure components, is presented by the system when components fail. To help reasoning about such systems, the use of formal methods becomes desirable. In previous work we introduced a graphical formal specification language, called Object-Based Graph Grammars (OBGG), for modelling asynchronous distributed systems. We also defined a method for automatically inserting classical fault behaviours into OBGG models. The obtained models could be analysed using simulation. In this paper a new method for automatically inserting fault behaviours into OBGG models, which is suitable for using verification as the analysis method, is proposed. Moreover, we show how to formally verify OBGG models in the presence of such faults. A two phase commit protocol is used to illustrate the contributions.

Keywords: Object-based graph grammars, distributed systems, fault-tolerance, model transformation, model checking.

1 Introduction

The development of fault-tolerant distributed systems is a difficult task. Besides dealing with the inherent complexity of concurrent systems, the developer also has to take distribution aspects into account and design the system in such a way that a well-defined failure behaviour, or the masking of failure components to users, is presented by the system when components fail [5]. In this context, the use of formal verification is an important analysis method because it allows

---

* This work is partially sponsored by IQ-MObile (CNPq/CNR) and DACHIA (FAPERGS/IB-BMBF) projects. Authors appear in alphabetical order.
** This author is partially sponsored by CAPES-Brazil.

C.A. Maziero et al. (Eds.): LADC 2005, LNCS 3747, pp. 80–100, 2005.

the developer to prove properties about the system. The verification method we consider in this work is model checking. The main advantage of using model checking for verifying fault-tolerant distributed systems relies on its exhaustive state-space checking procedure used to prove that a given property is true or not for a model of the system. Thus, using model checking one can reason about all interleaving possibilities of the fault occurrence w.r.t. the system behaviour.

In our activities we make use of a graphical formal specification language, called Object Based Graph Grammars (OBGG) [9], suitable for the specification of asynchronous distributed systems. Models defined with this formalism can be analysed using simulation [6] and verification (through model checking) [8,23]. There is also the possibility of generating code for execution in a real environment [6]. In [10] we proposed a method for automatically inserting (classical) fault behaviours, found in the literature of distributed systems, into OBGG models. According to the proposed method, the developer first defines a model M of the system under consideration and then selects the desired fault behaviour F that will be incorporated into M. A transformation of M, considering F, is performed and a new model M_F is obtained. M_F behaves as M in when the inserted fault F occurs in the system. In [10], we use simulation to analyse the behaviour of M_F. Once the developer is satisfied with the behaviour of M_F it is possible to generate code for execution in a real environment using M. In a real environment that exhibits the fault behaviour F, the system should behave as expected during the analysis phase.

Although simulation is an useful analysis method, especially for very large systems, the use of model checking allows one to check all the possible behaviours of the system. This characteristic is particularly useful for the analysis of fault-tolerant distributed systems, because developers are interested in guaranteeing a well defined behaviour for their systems in the occurrence of faults.

When using simulation for the analysis of OBGG models, the basic *synchrony* model (see Section 2) provided by the OBGG formalism is extended to specify minimum and maximum times for the reception of messages. Using this notion of time, developers can then specify timeouts for the reception of messages and detect the failure of components in the system. This idea is central for the automatic insertion of faults in OBGG models proposed in [10], because the failure detection is left for the developer to build (in terms of timeouts). Differently, in the approach based on verification presented here, we do not have this notion of time and the *synchrony* model is completely asynchronous. Thus, a new method from [10] is needed to provide developers with a basic failure detection mechanism. Using such mechanism, developers can work with a clear abstraction for the detection of faults in the system. They can specify fault-tolerant models that are analysable, after the automatic insertion of faults, using model checking. This aspect is further discussed in Section 4.

This way, in this paper we present two main contributions: (i) a new method, derived from [10], for automatically inserting fault behaviours into OBGG models that is suitable for using verification as the analysis method; and (ii) show how to formally verify OBGG models in the presence of such faults. As an example, we model a two phase commit protocol and discuss the verification of the protocol considering the crash fault model in different scenarios.

The paper is organized as follows: Section 2 presents the OBGG formalism and the model of a two phase commit protocol; Section 3 briefly reviews the verification approach and its application to the protocol; Section 4 presents the method for representing fault behaviours in OBGG models and the verification results for the protocol considering the crash fault behaviour; Section 5 covers related work and Section 6 presents conclusions and future works.

2   Object-Based Graph Grammars

The formal specification language used in this work is based on a restricted form of Graph Grammars (GG), called Object-Based Graph Grammars (OBGG). In this section we present an informal overview of GG and OBGG. The reader is referred to [22] and [9] for a more detailed presentation of these formalisms.

In a GG, the initial state of the system is called the initial graph. The state of the system is represented by a graph, called the system state graph. The system evolves through the successive application of rules that change the state of the system. A rule is composed of a left-hand side and a right-hand side, and can be applied whenever an image (match) of its left-hand side is sub-graph of the current system state graph. When applied, the rule brings the system to a new state defined as: items in the left-hand side not present in the right-hand side are deleted; items in the right-hand side not present in the left-hand side are created; and items present in both sides of the rule are preserved. Multiple rules can be applied in parallel if there is no conflict between them, i.e. do not modify (delete) the same item simultaneously. When a conflict situation exists, one of the candidate rules to be applied will be chosen in a non-deterministic way.

OBGG is a restricted form of GG w.r.t. the kinds of graphs and configuration of rules that are allowed. Like an object-based system, an OBGG model is composed of the definition of different classes. Each class is defined by a *type graph* and a set of rules. A *type graph* (see Figure 1) defines the attributes of the class and the messages that can be received by an object of that class. The rules (see Figures 2 and 3) are used to specify the behaviour of objects of that class upon the reception of messages defined in the *type graph*. The left-hand side of a rule always specifies a message being received by an object. At the right-hand side we specify the effect of applying the rule, which may be: changing attributes; creating new objects; and generating new messages. This way, the application of a rule may leave the system state graph in a configuration where various other matches may occur. The specification of a system where various classes are involved is given by the definition of each class complemented by an *initial graph* (see Figure 6) that contains instances of those classes in the initial state.

If we adopt the classification for message-passing models of distributed systems proposed in [17], the OBGG formalism provides: (*network topology*) directed graphs, where an object can only send messages to another if it has a reference to that object; (*synchrony*) completely asynchronous; (*failure model*) not defined, but introduced according to our approach – see section 4; (*message buffering*) infinite buffer of messages, that are received non-deterministically.

2.1 Two Phase Commit

In this section the two phase commit protocol described in [1] is modelled using the OBGG formalism. The protocol operates in rounds of communication and is composed of one coordinator and various participants. An important character-istic of the protocol is that a decision made during its execution must be taken by every participant.

In the first message round, the coordinator sends messages to participants in order to start the transaction. On receipt of this message, every participant votes to either commit or abort the transaction. The coordinator collects the votes sent to the participants and if: (i) all participants voted to commit, the transaction is committed; (ii) at least one participant voted to abort, the transaction is aborted; (iii) the reply of at least one participant is timed out, the transaction is also aborted. This concludes the first phase of the protocol. In the second phase, the coordinator sends to the participants the decision taken about the transaction, either commit or abort, depending on (i), (ii), and (iii).

The two phase commit protocol modelled in OBGG is composed of one co-ordinator and only two participants. This decision is due to the observation that the situations needed to prove properties of the protocol are covered by this configuration. With two participants we can represent the situations where: faulty and non faulty participants coexist; and only faulty or only non faulty participants coexist. Since the protocol is based on the votes of all participants, it is actually not important if one or various participants are faulty, because the transaction has to be aborted in such cases. Complementarily, the communica-tion among participants may take place during the recovery procedures. In these situations a participant tries to contact another participant in order to learn the outcome of the transaction. Again, we have the possibilities that either the con-tacted participant is available (at least one participant is available) or not (no participant is available)[1]. Moreover, the small number of objects in the model helps to decrease the possibility of state explosion during model checking.

The *type graphs* of *Coord* (representing the coordinator) and *Part* (repre-senting a participant) classes are shown in Figure 1[2]. Both *Coord* and *Part* may receive *Timeout* and *Recovery* messages. A *Timeout* message is used to repre-sent that an awaited response message will not arrive (will timeout), whereas a *Recovery* message is used to signal for the object that a previous fault situation in the object has ceased. As discussed in Section 4, the developer does not have to specify the generation of those messages, since they are part of the introduced fault behaviour, but he/she does have to specify the reaction of the model to these situations, e.g. what actions are taken when a response to a message is

---

[1] The protocol does not assure that a participant receives the outcome of a transaction if the other participants and the coordinator are faulty.

[2] The circle near the name of the class is a notation used to facilitate the reference of the class along the definition of the model (rules and *initial graph*).

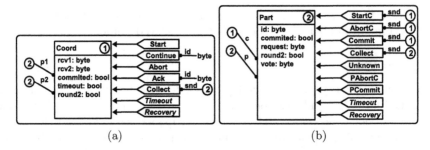

Fig. 1. Type graph of *Coord* (a) and *Part* (b) classes

timed out or when a fault in the object ceases. The use of those messages provides the developer with a clear abstraction regarding the detection of faults in the execution of the model.

In the following, the rules describing the behaviour of *Coord* and *Part* are discussed. Due to space constraints, not all rules are shown in Figures 2 and 3. However, the omitted rules are similar to the rules shown and will be discussed in the text. Basically, a *Start* message (rule *StartCommit*) is used to begin the execution of the protocol. This rule sends messages *StartC* to *Part* objects, requesting their votes. *Part* objects may non-deterministically answer with either *Continue* or *Abort* messages (rules *VoteForCommit* and *VoteForAbort*).

Upon receipt of a *Continue* message, *Coord* awaits for the vote of the *Part* objects (rule *RcvContinue1* and *RcvContinue2* (not shown, but similar to *RcvContinue1*, it handles the *Continue* message for the second *Part* object)). After receiving the *Continue* messages (all participants vote for commit), *Coord* enters the second phase of the protocol (*round2* attribute becomes *true*) and decides that the transaction must be commited. This way, a *Commit* message is sent to the *Part* objects (rule *DoCommit*). If the *Part* objects are able to complete the operation, an *Ack* message is sent to *Coord* (rule *Commit*). Analogously to the first phase, *Coord* awaits for *Ack* messages from *Part* objects (rules *RcvAck1* and *RcvAck2* (not shown, but similar to *RcvAck1*, it handles the *Ack* message for the second *Part* object)), ending the transaction (rule *Terminate*).

The behaviour discussed above shows the execution of a transaction where all *Part* objects vote to commit (*Continue* message) the transaction. If at least one of the *Part* objects votes to abort the transaction (rule *VoteForAbort*), then *Coord* decides to abort the transaction (rule *DoAbort* – not depicted, but similar to *DoCommit*, except that an *AbortC* message is sent instead of *Commit*). If other *Abort* or *Continue* messages are received after the decision to abort the transaction, they are ignored by *Coord* (rules *IgnoreContinue* and *IgnoreAbort* (not shown, but similar to *IgnoreContinue*, where the message *Abort* is ignored)). *Part* objects on receipt of an *Abort* message will abort the transaction (rule *Abort*) and send an *Ack* message to *Coord*. Finally, *Coord* awaits for *Ack* messages from *Part* objects (rule *RcvAck1* and rule *RcvAck2*) ending the transaction (rule *Terminate*).

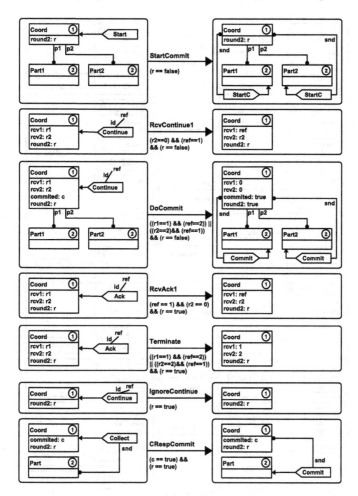

Fig. 2. Rules of *Coord* class

If a *Part* object becomes faulty during the transaction and does not know the final result of the transaction (attribute *round2* is *false*), it requests the result by sending a *Collect* message. Both *Coord* and *Part* objects can receive *Collect* messages (see rules *CRespCommit* and *CRespAbort* (not depicted, but similar to *CRespCommit*, where the *AbortC* message is generated) for *Coord*, and rules *PRespCommit* and *PRespAbort* (not shown, similar to *PRespCommit*, where the message *PAbortC* is generated) for *Part*) and reply with the final result of the transaction. Though, one particular situation occurs when a *Part* object requests the result from another *Part* object and it does not have the result, responding with an *Unknown* message to the sender (rule *PRespUnknown*).

Until now, we have explained the behaviour of the protocol without taking into account the possibility of faults from: (i) other objects, which are detected using the *Timeout* message; (ii) the object itself, when it recovers from a previ-

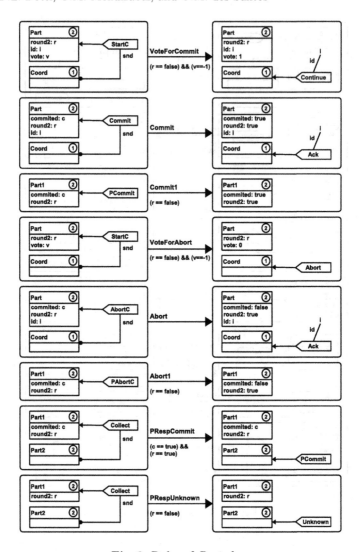

Fig. 3. Rules of *Part* class

ous fault, through the reception of a *Recovery* message. The rules used to model the fault-tolerant behaviour of both *Coord* and *Part* objects are presented, respectively, in Figures 4 and 5.

Because we are modelling asynchronous distributed systems, we need to make an object be able to detect if other object(s) is(are) faulty or not. If an object sends a message to another object and keeps awaiting for the response of that message, two possible outcomes can occur: (i) the correct response of the message; (ii) the receipt of a *Timeout* message, if the destination object is faulty. The reader should note that a *Timeout* message does not have parameters, and the object relies in its internal state in order to recognize the object to which the *Timeout* message belongs to.

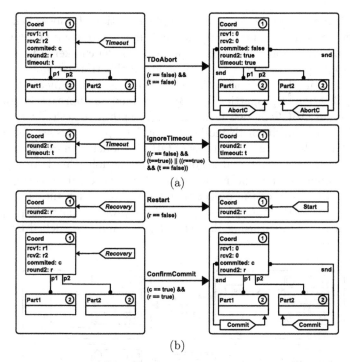

Fig. 4. Rules of *Coord* class (a) fault-tolerance and (b) recovery

The receipt of a *Timeout* message by *Coord* (see Figure 4) during the first phase results in the abortion of the transaction (rule *TDoAbort*). Besides, upcoming *Timeout* messages are ignored (rule *IgnoreTimeout*). Moreover, if *Coord* receives a *Recovery* message during the first phase it restarts the transaction (rule *Restart*), otherwise (if it receives it in the second phase) it resends the transaction result to all *Part* objects (rules *ConfirmCommit* and *ConfirmAbort* (not depicted, similar to *ConfirmCommit*, where a message *AbortC* is generated)).

During the recovery process, *Part* objects (see Figure 5) may request the result of the transaction from *Coord* object (rule *ReqResultCoord*) or ignore the *Recovery* message if it already knows the result of the transaction (rule *IsOk*). Thus, after requesting the transaction result from *Coord* object (*request* attribute is set to 1) and a *Timeout* message is received, the object requests the result of the transaction from another *Part* object (rule *ReqResultPart*) and ignore further *Timeout* messages (rule *ConsumeTimeout*).

Finally, Figure 6 presents an *initial graph* modelling an initial scenario which is used in Section 3.1 to illustrate the verification method. As explained before, the model is composed of one *Coord* object and two *Part* objects representing, respectively, the coordinator and two participants of the protocol.

3   Verifying OBGG Models

The use of model checking for the analysis of OBGG models has been introduced in [8]. The approach used for verifying OBGG is based on the translation of models. This aims to reuse the existing implementation of a model checker and take advantage of its enhancements. In particular, we translate OBGG models to PROMELA, the input language of the SPIN model checker [15]. The reader is referred to [8] for a more detailed discussion of this translation.

Complementary to the translation of OBGG models, a method is needed for the specification of properties. In [23] we defined an approach for specifying properties about OBGG models using Linear Temporal Logics (LTL) – the same temporal logic used in the SPIN model checker. LTL properties are defined using events produced during the executions of OBGG models. An event corresponds to the application of a rule, and is composed of the following information: the name of the applied rule; the name of the class whose rule was applied; and any internal attributes of the object that are necessary for the formula being specified. We follow the notion of events and the property patterns proposed in [3]. An event is expressed using the symbol ↑ *def* and is modelled by the LTL

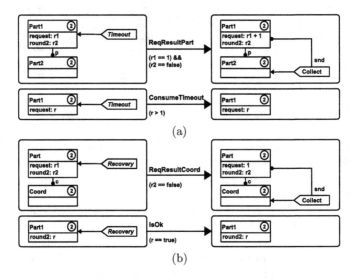

Fig. 5. Rules of *Part* class (a) fault-tolerance and (b) recovery

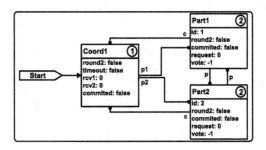

Fig. 6. Initial graph for two phase commit protocol

formula[3] (*!def* && *X def*). Although the *X* (next) temporal operator is used, the property patterns in use [3] are proven closed under stuttering and therefore can be analysed using the SPIN model checker.

Currently we have an environment [7] that enables a developer to graphically specify OBGG models and automatically translate them for verification. Moreover, our environment provides the generation of counter-examples in terms of OBGG abstractions, instead of the translated PROMELA model [23].

3.1   Verifying the Two Phase Commit

Now we discuss four properties that should be satisfied by our model of the protocol using the *initial graph* (initial scenario with one coordinator and two participants) of Figure 6:

(*i*) If a new transaction starts, eventually the coordinator will terminate it;
(*ii*) A participant always terminates a transaction (aborting or commiting);
(*iii*) If all participants vote for commit, they all will commit;
(*iv*) If at least one participant votes for abort, all participants will abort.

While (i) and (ii) prove the termination of both coordinator and participants, (iii) and (iv) ensure that *Part* objects take the same decision during a transaction. Proving these properties we achieve a high degree of confidence that the execution of a transaction is performed correctly. We need to specify these properties as LTL formulas in order to use model checking. In Table 1 we present the observed events of the model. The LTL formulas specified for the previously discussed properties are presented next. They follow the patterns defined in [3] and are closed under stuttering. For each formula, the pattern used is informed.

Table 1. Events used in LTL properties

Event	Object - Rule applied	Event	Object - Rule applied
(↑ *cs*)	*Coord - StartCommit*	(↑ *p1c*)	*Part1 - Commit*
(↑ *ct*)	*Coord - Terminate*	(↑ *p1c1*)	*Part1 - Commit1*
(↑ *ca*)	*Coord - DoAbort*	(↑ *p1a*)	*Part1 - Abort*
(↑ *cr*)	*Coord - Restart*	(↑ *p1a1*)	*Part1 - Abort1*
(↑ *cc*)	*Coord - DoCommit*	(↑ *pc*)	some participant - *Commit*
(↑ *cta*)	*Coord - TDoAbort*	(↑ *pc1*)	some participant - *Commit1*
(↑ *p1vc*)	*Part1 - VoteForCommit*	(↑ *pva*)	some participant - *VoteForAbort*
(↑ *p2vc*)	*Part2 - VoteForCommit*	(↑ *pa*)	some participant - *Abort*
		(↑ *pa1*)	some participant - *Abort1*

(*i*) If a new transaction starts (rule *StartCommit* of *Coord*) eventually the coordinator will terminate it (rule *Terminate* of *Coord*). The globally response pattern is used, resulting in the formula [] (↑ *cs* − > <> ↑ *ct*).
(*ii*) The participants always terminate. The following LTL formula (considering *Part1*) is actually proved for each participant:

---

[3] LTL formulas used in this paper follow SPIN syntax [15].

$<>$ ($\uparrow p1c \parallel \uparrow p1c1 \parallel \uparrow p1a \parallel \uparrow p1a1$). This formula uses the pattern of global existence. We do not use the pattern $[] <>$ because our model is finite and, therefore, it is not true that participants infinitely often terminate. Instead they eventually terminate in all possible finite executions.

(iii) To prove property (iii), we perform 2 steps:

(iii.a) Considering that all participants vote for commit, the coordinator will decide to commit the transaction: $<> \uparrow p1vc - > [] (\uparrow p2vc - > <> \uparrow cc)$. This formula uses the patterns global existence and response.

(iii.b) Once the coordinator decides to commit, no participant will abort. This formula uses the pattern absence after an event: $[] (\uparrow cc - > [] !(\uparrow pa \parallel \uparrow pa1))$. The rationale behind the above steps is that if the participants vote for commit, the coordinator will decide to commit, and the participants will not abort. Since the participants always terminate (formula (ii)), this means that they commit.

(iv) To prove property (iv), we also perform also 2 steps:

(iv.a) A coordinator may decide to abort for two reasons: the lack of a response message from a (faulty) participant - in this case a timeout occurs and the coordinator decides to abort (event $\uparrow cta$); or the participant votes for abort and then the coordinator decides to abort (event $\uparrow ca$). Here we prove that a vote for abort from a participant always precedes the decision to abort (not based in timeout), through the formula: $<> \uparrow ca - > !(! \uparrow ca \: U \: \uparrow pva)$. The existence before pattern was used.

(iv.b) If the coordinator aborts the transaction, the participants will not commit the transaction: $[] (\uparrow ca - > [] !(\uparrow pc \parallel \uparrow pc1))$. This formula uses the pattern absence after an event. The rationale is analogous to (iii). If one of the participants votes for abort, the coordinator will decide to abort. Since the participants always terminate (formula (ii)), it means that they abort the transaction.

All these formulas are valid for the scenario without faults. The results of the verification process are presented in scenario (a) of Table 2. As noted in the results, the generated state space of the model is small and the verification runs were due in almost no time. This occurs because in our model of the two phase protocol, only one transaction is executed and eventually terminates, defining a finite execution model for the problem (this model captures all the possible behaviours that are relevant to analyse the protocol). This may not be true for other models and in some cases an infinite behaviour may be required.

4 Representation of Fault Behaviours

As stated in [4] a system may change its state based on two event classes: normal system operation and fault occurrences. Based on this observation a fault can be modelled as a state transition of a system [13]. These transitions can be modelled through the use of additional virtual[4] variables, acting like guards to activate specific commands; and a group of guarded commands representing the specific fault behaviour, being activated whenever its associated guard is satisfied, by the assignment of a true value to it [12].

The addition of virtual variables and the notion of guarded commands can be viewed as a transformation of a model M into a model M_F that contains the behaviour of a selected fault behaviour F in its state space [13]. In [10], we adopted the same concepts to transform an OBGG model M into a model M_F that represents the system with the selected fault behaviour F. Due to the notion of time available in the simulation approach for OBGG (see Section 1), the transformations presented in [10] are not suitable for using verification as the analysis method. Therefore, now we extend the transformations of OBGG models [10], with a basic failure detection mechanism which makes it suitable for using verification as the analysis method.

In the OBGG formalism, this model transformation corresponds to the addition of guarded commands that either trigger the fault behaviour or not. This way, in OBGG the left-hand side of a rule corresponds to the guard of the command and the right-hand side of the rule (application of the rule) corresponds to the execution of the guarded command. We model the fault behaviour F in an OBGG model inserting virtual variables (used for the guards) and messages (used for the activation/deactivation of the fault behaviour) in every (*type graph*) class of the model. Besides, we need to create and change all the rules defined for the classes that appear in the model. Depending on the fault behaviour F, different rule transformations may occur.

In order to activate the fault behaviour for an object we insert a special message, in the *initial graph* (initial configuration) of the OBGG model, addressed to the object we are interested to activate the fault behaviour. This approach is used because messages in OBGG are received non-deterministically. The non-determinism ensures that all the possible combinations for the activation of the fault behaviour for an object will be analysed during verification, i.e. the deletion of the message activating the fault behaviour is arbitrary. The same idea of inserting a message to activate the fault behaviour is used to de-activate the fault behaviour and enable the recovery of the object from the fault behaviour. This de-activation of the fault behaviour is characterized with the generation of a *Recovery* message (see Section 2.1) to the object recovering from the fault.

When selecting the fault behaviours to be represented we have adopted the classification found in [12]. There, fault behaviours are classified into the following categories: fail-stop, crash, receive omission, send mission, general omission, and Byzantine. From these fault behaviours, we do not model the fail-stop and Byzantine behaviours. The reason we do not model the fail-stop behaviour is that we consider it to be to restrictive and that the crash behaviour is more commonly used in the literature of distributed systems. In the case of Byzantine faults, it is subject of future work.

---

[4] The term virtual is used to qualify variables that are not part of the desired system itself, but part of the introduced fault behaviour.

4.1   Crash Fault Behaviour

In the crash fault behaviour a process fails by halting. The processes that maintain communication with the halted process are not warned about the fault. Figure 7 shows an algorithm to transform an OBGG model M (without fault behaviour) into a model M_F that incorporates the behaviour of a crash fault F_C. In order to add the behaviour of a crash fault (Figure 7) the transformation procedure inserts a virtual logical variable in every *type graph* of the classes of the model. Depending on the value of this variable, the object may exhibit the fault behaviour (*down* is *true*) or not (*down* is *false*). Besides, new rules are added to activate the fault behaviour and to cease it. To illustrate, the rules *Crash* and *Uncrash* for the *Part* class are presented in Figure 8.

As defined in Figure 7, we also have the addition of new rules to represent the fault behaviour in the model. These rules represent the behaviour of a crash fault upon the reception of each message. The behaviour consists in consuming the message, making no modification to the internal state of the object, creating no other object(s), and generating no message(s).

```
1    For every class in the model {
2        // guard
3        in the type graph:
4            insert a boolean "down" variable
5        in the initial graph:
6            set "down" to "false" (used as the guard)
7
8        // activation and deactivation rules
9        create a new rule called "Crash"
10       insert as the activation message a "Crash" message
11       if "down" is "false"
12           this rule sets the guard "down" to "true"
13
14       create a new rule called "Uncrash"
15       insert as the activation message an "Uncrash" message
16       if "down" is "true"
17           this rule sets the guard "down" to "false"
18
19       // create rules with the behaviour in a crash situation
20       For all rules in the class definition {
21           replicate the current rule and for each replica
22               insert a guard "down: true"
23           modify the right-hand side:
24                   the internal attributes remain unchanged
25                   no messages are generated
26                   no objects are created
27               if the received message had a "snd" (sender) parameter
28                   send a "Timeout" message to the sender "snd"
29       }
30
31       // change original rules to work only if not crashed
32       For all original rules (not replicas) in the class definition {
33           insert a guard "down: false"
34       }
35   }
```

Fig. 7. Algorithmic transformation over a model to represent a crash fault behaviour

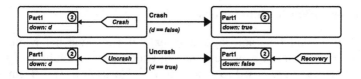

Fig. 8. Rules *Crash* and *Uncrash* used to activate/deactivate the crash fault behaviour

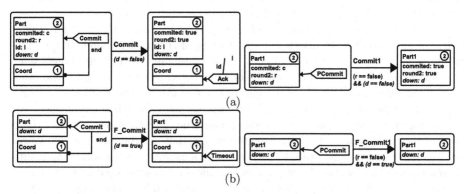

Fig. 9. (a) Rules without fault behaviour and (b) rules considering fault behaviour

According to [16], the internal state of a crashed process is undefined – when the process recovers from the fault its state is re-initialized. However, in [12] the representation of a crashed process is that the internal state remains unchanged (the last state of the process is the same state when recovered from the fault). In fact, the representation of crash presented in this work adopts the last definition. Mainly because it is useful to the case study: the recovery process of the two phase commit protocol relies on persistent internal state of the isolated members. Nevertheless, the first definition could be modelled using a *Crash* message that re-initializes the object when the fault occurs. This way, the application has to manage itself to update the object, when it recovers from the fault, with respect to the current state of the distributed system.

Though, in order to detect a faulty object in the model it is necessary to add timeouts to the messages sent to the faulty object. Since OBGG models used for verification are completely asynchronous, we have to explicitly represent the lack of a response as a message that is generated when a fault behaviour is activated. Thus, messages that have a reply associated, when processed by a faulty object, generate a *Timeout* message in reply to the sender (*snd*) which is assumed as a parameter of the incoming message (*snd*) (see Figure 9, rule *Commit* and *F_Commit*). On the other side, messages that do not have reply associated are simply consumed by the crashed object without generating any reply (see Figure 9, rule *Commit1* and *F_Commit1*).

4.2   Omission Fault Behaviours

A process in a send omission fault may exhibit the same behaviour as in a crash fault. Furthermore, a process may fail to transmit a subset of the total messages that it was supposed to send [14]. An algorithm to transform a model M (without fault behaviour) into a model M_F that incorporates the behaviour of a send omission fault F_{SO} is presented in [10]. Due to space constraints we do not present the revisited algorithm but we explain the main idea behind the transformation used to represent this fault behaviour.

In order to add the behaviour of a receive omission fault, a virtual logical variable ($rcv\_omitted$) is inserted in every class of the model. Depending on the value of this variable, the object may exhibit the fault behaviour (true value) or not (false value). New rules are created to activate and deactivate the fault behaviour. Moreover, modified replicas of the already defined rules are also created. Those modified replicas are used to represent the fault behaviour and have in their guards the $rcv\_omitted$ variable set to true. The right hand side of these rules specify that no objects are created or internal attributes changed. Moreover, no messages are generated. Instead, when a message with a sender attribute is received in the left hand side, a *Timeout* message is generated to that sender in the right hand side (analogous to the *Crash* fault model). The original rules of a class are not modified (guards are not inserted), since in the receive omission fault behaviour a process may fail to receive only a subset of messages. That is why we do not insert a guard on the original rules, allowing the choice of a rule to be applied in a non-deterministic way (once the guard is *true*).

The send omission fault behaviour is analogous to the receive omission fault behaviour. The main difference is that in this fault behaviour only a subset of the total messages sent by the fault process are actually sent. Though, it can be modelled in a similar way to the receive omission fault behaviour. Moreover, the general omission behaviour specifies that a process may experience both send and receive omissions [20]. Using these concepts we model the general omission as a semantic composition of the send and receive omission fault behaviours.

4.3   Verifying the Two Phase Commit with Crash Fault Behaviour

In order to apply model checking to a given system specified in OBGG, considering the presence of a given fault behaviour, we have to: (i) choose which fault behaviour F we want to reason about; (ii) transform the OBGG model M according to the algorithmic description of the fault; (iii) translate the transformed OBGG model M_F to the input language of the model checker. After that, we need to describe properties to be checked using LTL, start the model checking tool, and analyse the output results.

We carried out these basic steps for the two phase commit protocol: (i) the crash fault behaviour was chosen; (ii) the original OBGG model, presented in Section 2.1 was transformed into another model according to the algorithm shown in Section 4.1; (iii) the resulting model was translated to PROMELA using our tool [7]. After that, we extended the verifications shown in Section 3.1

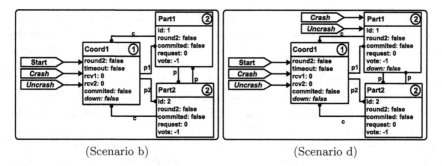

(Scenario b)   (Scenario d)

Fig. 10. Example of initial graphs for two phase commit protocol considering faults

Table 2. Verification results for scenarios (a) to (e) and formulas (i) to (iv)

Sc.	Resources	For.(i)	For.(ii)	For.(iii.a)	For.(iii.b)	For.(iv.a)	For.(iv.b)
(a)	States	17838	1416	5753	3055	1497	24662
	Time (m:s)	≈ 00:00	≈ 00:00	≈ 00:00	≈ 00:00	≈ 00:01	≈ 00:01
	Memory	3.032MB	2.724MB	2.929MB	2.827MB	2.724MB	3.236MB
(b)	States	805303	72508	551527	280820	74261	1.78212e+06
	Time (m:s)	≈ 02:16	≈ 00:12	≈ 01:06	≈ 00:30	≈ 00:06	≈ 03:48
	Memory	30.065MB	6.616MB	23.921MB	15.217MB	5.284MB	57.201MB
(c)	States	249400	47828	157446	87653	25477	553126
	Time (m:s)	≈ 00:34	≈ 00:07	≈ 00:12	≈ 00:06	≈ 00:01	≈ 00:55
	Memory	11.224MB	4.977MB	8.971MB	6.718MB	3.544MB	19.723MB
(d)	States	9.19774e+06	1.3985e+06	1.88823e+07	3.1813e+06	554593	2.09131e+07
	Time (m:s)	≈ 17:30	≈ 03:22	≈ 01:07	≈ 04:08	≈ 00:29	≈ 38:38
	Memory	319.038MB	71.230MB	609.137MB	139.224MB	20.747MB	667.096MB
(e)	States	-	9.59151e+06	1.12832e+07	2.50637e+07	8.84848e+06	9.26983e+07
	Time (h:m)	-	≈ 00:59	≈ 00:41	≈ 01:49	≈ 00:35	≈ 50:16
	Memory	-	454.206MB	466.904MB	1071.166MB	298.046MB	3033.150MB

to consider the inserted crash fault behaviour. The verification scenarios were: (a): without faults; (b): only *Coord* fails and recovers; (c): only one *Part* fails and recovers; (d): both *Part*s fail and recover; (e): *Coord* and one *Part* fail and recover. These scenarios were implemented in the model by creating new *initial graphs* with the addition of messages used to activate/deactivate the crash fault behaviour of the desired objects, as exemplified in Figure 10.

The reader should note that the non-determinism found in the message reception of the OBGG formalism obliges the verification process to traverse all the possible interleaving between the processing of the *Crash* message and other messages, that may be present (at any time) in the model. The inclusion of both *Crash* and *Uncrash* messages in the *initial graph* are possible, since the *Uncrash* message can only be applied after the consumption of the *Crash* message.

For scenarios (a), (b), (c) and (d), all LTL formulas proposed in Section 3.1 are valid and yield the results shown in Table 2. The formulas were verified in an Intel Xeon 2.2 GHz Processor with 4 GB of memory allocated to the SPIN model checker. As expected, the addition of the fault behaviour increases the size of the model considerably and its verification time.

Now we discuss scenario (e), where *Coord* and a *Part* are faulty. With exception of LTL formulas (i) and (iii.a), all the other formulas are valid also for scenario (e). Formula (i) does not verify. This is a possible outcome of the basic protocol that the verification is showing. There is the possibility that *Coord* does not terminate the transaction (rule *Terminate*) because the faulty *Part* may already have received the decision (*DoCommit* or *DoAbort*) and acknowledged it, but in the meanwhile *Coord* may have failed and then looses the acknowledgement. When *Coord* recovers, it tries to repeat the second phase but then *Part* may have failed and will not respond and *Coord* will continue with a pending transaction. The faulty *Part* will then recover and assume the persistent state, i.e. transaction finished[5]. The mechanism proposed in the literature [1] to deal with this is to extend the protocol with a garbage collection mechanism whereby *Coord* tries to resolve pending transactions repeating the second phase.

Continuing the discussion of scenario (e), since formula (ii) can be verified, we know that all participants reach the termination of the protocol, although the coordinator may not. Given this, we proceed to prove that the participants finish the transaction with coherence, with properties (iii) and (iv). Formulas (iii.b), (iv.a) and (iv.b) are also valid for scenario (e), but (iii.a) does not verify. This is also an expected behaviour and it happens because a participant may vote for commit while the coordinator is faulty. The recovery action of the coordinator will ask for a new vote and this time the participant may have failed. The absence of the participants vote will generate a timeout and the coordinator will decide to abort. To prove that when both participants vote for commit the coordinator will decide for commit we have to filter the behaviour of the coordinator to the traces where it does not restart the voting procedure (to avoid the above described possible behaviour). The resulting formula is: $[]! \uparrow cr - > (<>\uparrow p1vc - > [](\uparrow p2vc - > <> \uparrow cc))$. The results for formula (iii.a) in scenario (e) of Table 2 consider this formula.

The practical limitations we faced during our work are strongly related to the state space explosion problem, which is a general problem when using model-checking. The verifications reported in Table 2 make use of the same SPIN options. Among the options we choose compression in order to save memory - our machine is limited to 4GB of memory and formula (iv.b) used more than 3GB for scenario (e). The drawback of this option is that the processing time is penalized. The reader should note that in scenario (e) of Table 2 the timescale is "h:m" and not "m:s" as for the other scenarios.

Further reporting about our experiences in this sense, we have modelled the scenario where all instances fail. The verification of this case was not possible due to insufficient memory. The fast increase in the state space is understandable since the additional behaviours have to be considered in all possible interleaving with the already existing behaviour. For instance, in the situation where all instances fail, if compared to scenario (e), the model checking procedure would have to consider all possible interleaving of the crash of the second participant with respect to the possible computations of scenario (e).

With the above discussed experiments, we have exemplified our approach and verified various important properties of the two phase commit protocol.

---

[5] Due to space restrictions we do not present this graphical counter-example.

During the verification process our model was naturally extended and corrected. Extensions could be proposed to the verification here proposed. For instance, it would be possible to investigate the situation where a selected object repeatedly fails. In this case the *initial graph* would have several *Crash* and *Uncrash* messages posted to the selected object. *Crash* messages are not consumed during a crash and therefore continue in the state graph until the recipient object could consume it, i.e. the object is not crashed. For the specific case of the two phase commit protocol, since the recovery actions are atomic, the successive application of crash/recovery to an object would lead to the repetition of states.

Although we used the same formulas defined in Section 3.1, one can specify and verify other properties that consider the fault behaviour by selecting a specific set of computations of the model involving faults. This way, it is possible to specify properties (via LTL formulas) that consider the activation and deactivation of the fault behaviour (e.g. *Crash* rule and rules defined for consuming the *Recovery* message, respectively) as part of the property specification.

5   Related Work

Concerning related work, we have mainly surveyed the literature trying to identify approaches that allow developers to reason about distributed systems using formal verification (model checking) in the presence of faults. Complementarily, related work on model checking Graph Grammars (GG) are also reported.

The SPIN model checker [15] enables a developer to specify a system using the formal language PROMELA, and specify properties using LTL. The similarity is that SPIN allows one to verify models using channel abstractions that may loose messages. This feature is provided by the supporting tool. Other fault behaviours are not provided. Thus SPIN provides message losses through the run-time environment. Faults are not represented as part of the system model. Since in this work we embed the fault behaviour in the system model, we can use different tools to analyse the effects of the fault in the model. In [10] we used simulation and in this paper we introduce model checking as analysis tool.

Another work, which is directed to the development of mobile systems, is presented in [11]. The Distributed Join Calculus is a process calculus that introduces the notions of locations and agents, where locations may be nested. The calculus provides the notion of crash of a location, whereby all locations and agents in the crashed location are halted. A crash of a location changes the behaviour of basic characteristics hindering any kind of reaction, like for communication or process migration. Systems that are represented with this calculus can be verified semi-automatically using theorem proving.

In [19] an extension to the basic I/O Automata [18] model includes fault behaviour in critical and distributed systems. According to this extension, actions are classified in normal, fault and recovery actions. Each class of actions can be active or not, defining for each instant a set of actions that can be taken. Thus, it is possible for fault actions to disable some normal actions, in accordance with the fault class modelling, and recovery actions to activate some normal actions again. Altogether, [19] offers abstractions (classes of actions and activation/deactivation lists) such that the developer can represent the fault behaviour of interest. In our work, fault representation is done by transforming the specification automatically. Moreover, although there is tool support for the verification of I/O Automata, it is not clear whether this extension is also supported.

The work of [24] presents a proposal very similar to this one. A language based in guarded commands is used to model systems. This language is mapped to the input language of the SMV model checker [2] for verification. The paper does not mention the semantic compatibility of the original model and the one generated as input to SMV. The representation of fault behaviours uses the same underlying ideas as this paper. However, the approach is not based on a transformation of the original model but the developer must explicitly describe the desired fault behaviour.

Complementarily, there are few contributions addressing the verification of GG. Some of these contributions propose approaches to the verification of infinite-state and others finite-state GG. The work in this paper uses a finite-state approach. Among these, main related work are CheckVML and GROOVE which are described and compared in [21]. A key difference from those contributions to OBGG model checking is that the later allows dealing only with a restricted form of GG (OBGG), while CheckVML and GROOVE address GG in general. On the one side, OBGG imposes important restrictions, but on the other side the restrictions fits well in the object-based style and reduce the problem of finding matches of rules in the current system state graph. Another difference is that [21] focus on reachability properties while in OBGG model checking allows one to state various kinds of properties.

6 Final Remarks

In this paper we presented a method for the specification and analysis (using model checking), of fault-tolerant distributed systems modelled with the Object-Based Graph Grammars (OBGG) formalism. We considered the definitions of crash and omission fault behaviours and how to automatically introduce them into an OBGG model. This transformation leads to a new model that incorporates the fault behaviour. Special messages are used to activate/deactivate the fault behaviour. When activated, the fault behaviour generates special events (*Timeout* and *Recovery* messages) that the model should respond to in order to deal with the faults. As discussed in this paper, the use of such events provide a well defined abstraction w.r.t. the fault behaviour that developers can use to model and analyse (using model checking) fault-tolerant distributed systems.

In particular, two main characteristics of the OBGG formalism made it suitable for the approach: (i) to be based on rules – making it easy to add guards for the activation/deactivation of the introduced fault behaviour; (ii) non-deterministic reception of messages – enabling the analysis of all possible combinations of the activation/deactivation of the fault behaviour. Since models described in OBGG can be formally verified using model checking, as well as simulated, we provide methods and tools to help reasoning about distributed systems in the presence of faults.

We have focused our activities in the Crash fault model. This is because it is often used in the literature. Moreover, we could satisfactorily illustrate the general idea of handling with fault models using it together with the case study. It is possible to envision the automatic modification of messages in order to model random corruption. However, a deeper study is needed to identify the possibility of modelling malicious intentions of messages.

Although the case study was small in terms of instances, we could prove important properties. As discussed in Section 4.3, even this small case lead to considerable resource consumption. State space explosion is often a problem when using model checking techniques and various techniques can be employed to reduce it. In our case we are translating from OBGG to the input language of a model checker. This translation introduces basic structures and processes to represent the OBGG abstractions that are costly in terms of time and space. An important future work comprises the optimization of this translation.

References

1. K. P. Birman. *Building secure and reliable network applications.* Manning Publications, USA, 1996.
2. J. R. Burch et al. Symbolic model checking: 10^{20} states and beyond. *Information and Computation*, 98(2):142–170, 1992.
3. M. Chechik and D. O. Păun. Events in property patterns. In *5th and 6th Int. SPIN Workshops*, volume 1680 of *LNCS*, pages 154–167, Germany, 1999. Springer-Verlag.
4. F. Cristian. A rigorous approach to fault-tolerant programming. *IEEE Trans. on Soft. Eng.*, 11(1):23–31, 1985.
5. F. Cristian. Understanding fault-tolerant distributed systems. *Communications of the ACM*, 34(2):56–78, 1991.
6. F. L. Dotti, L. M. Duarte, B. Copstein, and L. Ribeiro. Simulation of mobile applications. In *2002 Communication Networks and Distributed Systems Modeling and Simulation Conference*, pages 261–267, USA, 2002. SCS.
7. F. L. Dotti et al. An environment for the development of concurrent object-based applications. *Eletronic Notes in Theoretical Computer Science*, 127:3–13, 2005.
8. F. L. Dotti, L. Foss, L. Ribeiro, and O. M. Santos. Verification of object-based distributed systems. In *6th Int. Conference on Formal Methods for Open Object-Based Distributed Systems*, volume 2884 of *LNCS*, pages 261–275, France, 2003. Springer-Verlag.
9. F. L. Dotti and L. Ribeiro. Specification of mobile code systems using graph grammars. In *4th Int. Conference on Formal Methods for Open Object-Based Distributed Systems*, volume 177 of *IFIP Conference Proceedings*, pages 45–64, USA, 2000. Kluwer Academic Publishers.
10. F. L. Dotti, O. M. Santos, and E. T. Rdel. On the use of formal specifications to analyze fault behaviors of distributed systems. In *1st Latin-American Symposium on Dependable Computing*, volume 2847 of *LNCS*, pages 341–360, Brazil, 2003. Springer-Verlag.
11. C. Fournet et al. A calculus of mobile agents. In *7th Int. Conference on Concurrency Theory*, volume 1119 of *LNCS*, pages 406–421, Italy, 1996. Springer-Verlag.

12. F. C. Gärtner. Specification for fault tolerance: a comedy of failures. Technical Report TUD-BS-1998-03, Department of Computer Science - Darmstadt University of Technology, Germany, 1998.
13. F. C. Gärtner. Fundamentals of fault-tolerant distributed computing in asynchronous environments. *ACM Computing Surveys*, 31(1):1–26, 1999.
14. V. Hadzilacos and S. Toueg. A modular approach to fault-tolerant broadcasts and related problems. Technical Report TR94-1425, Department of Computer Science, Cornell University, USA, 1994.
15. G. J. Holzmann. The model checker SPIN. *IEEE Trans. on Soft. Eng.*, 23(5):279–295, 1997.
16. P. Jalote. *Fault tolerance in distributed systems*. Prentice Hall, USA, 1994.
17. L. Lamport and N. Lynch. Distributed computing: models and methods. In *Handbook of theoretical computer science*, volume B: formal models and semantics. Elsevier, 1990.
18. N. A. Lynch. *Distributed algorithms*. Morgan Kaufmann, USA, 1996.
19. T. S. Perraju, S. P. Rana, and S. P. Sarkar. Specifying fault tolerance in mission critical systems. In *1st IEEE High Assurance Systems Engineering Workshop*, pages 24–31, Canada, 1996. IEEE Computer Society Press.
20. K. J. Perry and S. Toueg. Distributed agreement in the presence of processor and communication faults. *IEEE Trans. on Soft. Eng.*, 12(3):477–482, 1986.
21. A. Rensink, Á. Schmidt, and D. Varró. Model checking graph transformations: a comparison of two approaches. In *Int. Conference on Graph Transformation*, volume 3256 of *LNCS*, pages 226–241. Springer-Verlag, 2004.
22. G. Rozenberg, editor. *Handbook of graph grammars and computing by graph transformation*, volume 1: Foundations. World Scientific Publisher, 1997.
23. O. M Santos, F. L. Dotti, and L. Ribeiro. Verifying object-based graph grammars. *Eletronic Notes in Theoretical Computer Science*, 109:125–136, 2004.
24. T. Yokogawa, T. Tsuchiya, and T. Kikuno. Automatic verification of fault tolerance using model checking. In *2001 Pacific Rim Int. Symposium on Dependable Computing*, pages 95–102, Korea, 2001. IEEE Computer Society Press.

The Zerberus Language: Describing the Functional Model of Dependable Real-Time Systems

Christian Buckl, Alois Knoll, and Gerhard Schrott

TU München, 85748 Garching b. München, Germany
buckl@in.tum.de
http://www6.in.tum.de

Abstract. A growing number of safety-critical systems is controlled by computer systems. Currently these systems are often built from scratch. The Zerberus System assists the developer in the design and implementation process. Main features of the Zerberus System are generality, dependability, real-time predictability, the ability to be certified and cost-efficiency.

The main concept of the Zerberus System is the platform independent specification of the functional model by the developer. The functional model specifies the functional elements (tasks), the relation between these elements, the interaction of the system with the environment and the temporal constraints. On the base of the functional model the Zerberus System automatically generates the fault-tolerance mechanisms. Thus the task of the developer is restricted to the implementation of the application-dependent code.

In this paper we present one major part of the Zerberus System: the Zerberus Language that is used to specify the functional model of the control applications.

1 Introduction

Many safety-critical control systems are automated by the use of computer systems. Although the main fault-tolerance mechanisms are known for a long time [1,2] a general approach in the sense of reusing fault-tolerance mechanisms is missing. Most systems are therefore built from scratch and the application functionality is mixed with the fault-tolerance mechanisms. This leads to a time-consuming and cost-intensive development process.

Within the Zerberus System a development process is suggested to the user that attempts to reduce the development times and costs, while increasing the reliability and safety of the software. The Zerberus System emphasizes five different features: generality (by supporting the development of computer systems for various applications and domains, e.g. space, medical and traffic engineering), dependability (by providing fault-tolerance mechanisms to comply with the safety and availability requirements), real-time capability (by enabling the satisfaction of hard real-time constraints), the ability to be certified (by meeting

C.A. Maziero et al. (Eds.): LADC 2005, LNCS 3747, pp. 101–120, 2005.

certification standards e.g. DO-178B [3], IEC 61508 [4] and assisting the certification process by the system's architecture) and cost-efficiency (by supporting commercial-of-the-shelf (COTS) hardware and by accelerating the development process).

The main concept of the Zerberus System to achieve these features is to separate the functional design of the application from the platform dependent implementation and to provide a set of pre-implemented fault-tolerance mechanisms. This separation is realized by the specification of the functional model of the application. This model specifies the functional elements, the relation between the elements, the interaction of the system with the environment and the temporal constraints. On the basis of the functional model the Zerberus System is enabled to generate automatically the necessary fault-tolerance mechanisms. Thus the task of the developer can be minimized to the implementation of the application-dependent code. The automatic code generation of the fault-tolerance mechanisms is performed by using templates that are implemented independent from a certain application. The templates are carefully designed and coded and we intend to obtain a certification for these templates from the German certification authority TÜV. By reusing certified templates for the fault-tolerance mechanisms the development process can be accelerated and the error rates in comparison to a repeated reimplementation of these mechanisms can be reduced.

The fault-tolerance mechanisms that are currently supported are based on structural redundancy. At least three redundant units are executed in parallel. In the following we denote the redundant units as Zerberus units. The system offers facilities for synchronization, voting, exclusion of erroneous units and reintegration of repaired Zerberus units.

In this paper we focus on the Zerberus Language. This language is used for the specification of the functional model by the developer. The language had to be designed in a way that the fault-tolerance mechanisms could be realized based on this model. Therefore the main goals for the language were the suitability for replica determinism (to enable a comparison of the states of the redundant Zerberus units for an error detection) and the existence of previously known points in time for voting (to enable the implementation of distributed voting and synchronization algorithms).

The paper is structured as follows: section. 2 discusses related work, section. 3 introduces the development process proposed by the Zerberus System to clarify the role of the Zerberus Language. In addition the requirements on the language are elaborated. The main concepts of the Zerberus Language are then described in section. 4 in an informal way, while the exact semantics are specified in section. 5. At the end of the paper the concrete syntax of the Zerberus Language is pointed out for a concrete control program in section. 6 and the work is summarized in section. 7.

2 Related Work

Different research groups have observed the demand for a development process for safety critical real-time systems. Most of these solutions are based on the time-triggered paradigm [5]. The time-triggered approach guarantees one important aspect that is absolutely necessary for fault-tolerance mechanisms: determinism.

One important representative for the time-triggered approach is TTP/C [6]. TTP/C, the Time-Triggered Protocol, is a TDMA protocol designed to handle highly dependent real-time applications implemented in distributed networks. The protocol offers clock synchronization, clique avoidance, deterministic message sending and membership services [7]. The TTP/C protocol itself offers nevertheless no built-in fault-tolerance mechanisms at application level. Several other projects addressed this problem (MARS [8] or DECOS [9]). All these approaches have one major drawback in our opinion: the restriction to special hardware (like TTP/C controllers), programming languages or operating systems.

Our attempt was to design a development process that allows the usage of commercial-off-the-shelf hardware and that has no constraints towards programming languages and operating systems. This approach is shared with the research project Giotto [10,11], from the University of California at Berkeley. On the one hand, Giotto is based on the time-triggered approach, but on the other hand it also uses results of the research on synchronous languages like Esterel [12] or Lustre [13]. Like the synchronous languages Giotto introduces an abstraction level that separates the software design process from the actual hardware. By using the concept of FLET (Fixed Logical Execution Times), the applications designed with Giotto are not only deterministic regarding the values of the results (like Esterel, Lustre), but also have a deterministic temporal behavior. Thereto Giotto offers a language for the specification of the platform independent functional model for distributed real-time applications. A platform in the sense of Giotto (and in the sense of Zerberus) comprises the hardware, the operating system and the programming language. The mapping of the platform independent functional model to executable code is realized by a code generator. Since Giotto was designed primarily for the use in distributed systems Giotto has no built-in fault-tolerance.Within our project we developed the Zerberus Language, which is based on Giotto, to describe the functional model of the safety critical system.

Another tool intended for modeling and implementation of embedded systems is TIMES [14]. Within TIMES the developer models a system and the abstract behavior of the environment. By using a simulator the user can validate the dynamic behavior and verify the schedulability [15] of the system. A code generator for the synthesis of C-code on a LegoOS platform is provided. Like Giotto the tool TIMES was not intended for the use for dependable systems.

Several goals of the Zerberus System are also shared with Erlang [16,17]. Erlang is a programming language designed for programming real-time control systems. The language offers many features that are more commonly associated with an operating system than a programming language like concurrent process, scheduling or garbage collection. Fault-tolerance, fail-over, take-over is built right into the platform and concurrent processing is one of its strengths. In contrast to the Zerberus System, Erlang was designed only for soft real-time systems. Another difference is the programming extent: while Erlang is used for implementing the whole application, the Zerberus Language is only used for the specification of the functional model. For the implementation of the pure application code the developer can use a common, familiar programming language like C.

3   Development Process

The Zerberus System suggests different steps in the development process for dependable systems. In each step the system assists the developer to accelerate the process (for example by automatic code generation) and to improve the results by tool support or by providing guidelines. The individual steps to produce executable code are illustrated in fig. 1 and are described below. Since for most of the safety-critical systems a certification by an authority is required this problem is also addressed.

The description of the individual steps is focused on the requirements towards the Zerberus Language.

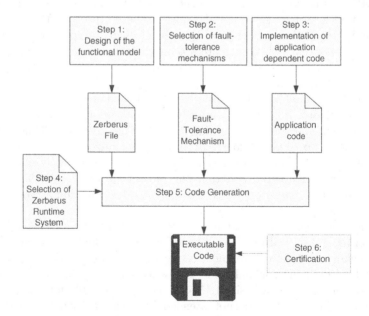

Fig. 1. Development process

3.1   Specification of the Functional Model

Within this step the user has to specify the functional elements of the application, their relationship towards each other and to the environment as well as the temporal constraints. The specification is realized by the use of the Zerberus Language. Since the specification of the functional model should be independent of a specific platform, the Zerberus Language has to be designed in a way to support this independency. A platform in the context of the Zerberus System is understood as the hardware, the operating system and the programming language.

The Zerberus Language was designed very simple and intuitive to avoid an error source and a long-lasting learning process. The language is not based on a certain programming language or operating system to comply with the generality requirement of the Zerberus approach.

3.2   Analysis of the Requirements on the Dependability

Currently the Zerberus System offers active structural redundancy as fault-tolerance mechanism. At least three Zerberus units compute the application in parallel. At specified points in time the units perform a distributed voting and synchronization algorithm. Erroneous units are excluded from the computation and can perform error recovery algorithms. Since error recovery algorithms are in most times application dependent the current run-time systems offer only a restart of the system or a reboot. In addition the developer can specify further fault reactions and recovery algorithms. After a successful completion, protocols allow the reintegration into the running system.

Since a replication of identical units allows no toleration of design errors, the system also supports diversity of hardware and software. While hardware diversity leads to no or only few additional costs as a result support of COTS hardware, N-Version programming is often not considered due to the extra effort necessary for the implementation of the individual versions.

As a result of these considerations several requirements are posed to the Zerberus Language. First of all the language must support the replica determinism: during the system execution it must be possible to compare the states of the individual Zerberus units for error detection. Especially due to the support of N-Version programming this is not a trivial requirement.

Another requirement that arises due to the voting is the existence of deterministic points in time when the voting should be performed. The existence of deterministic points in time is on the one hand the main requirement to allow the implementation of a distributed voting algorithm, on the other hand it also allows the implementation of a distributed clock synchronization algorithm.

The voting in the Zerberus system is performed in two rounds to additionally support the usage of a non-reliable communication network and is based on the voting algorithms as suggested by Klaus Echtle [18]. The voting messages are also used for the synchronization algorithm [19,20,21]: by means of the expected and the actual arrival time of the voting messages a logical global clock can be computed. The initial clock synchronization at start up is based on the algorithm implemented in the TTP/C [6] protocol.

To support the re-integration of a previously excluded Zerberus unit, the system must offer facilities for state synchronization. Since the algorithms are realized automatically by the system a derivation of the state of the individual units must be possible out of the functional model.

Finally, in order to achieve a reduction of the implementation effort for N-Version programming the code that has to be implemented by the developer should be restricted to the pure application dependent code.

3.3   Implementation of Application Dependent Code

In this step the developer has to implement code for the application. As already implied in the previous section this code is restricted to the pure application dependent functionality of the main parts which were identified within the design process of the formal model. By this restriction, the implementation effort can be reduced to a minimum.

The implementation step is platform dependent. This implies that for every platform used, the code has to be reimplemented by the developer.

3.4   Selection of Run-Time Systems

Run-time systems realize the execution of the application on the individual platform and provide the fault-tolerance mechanisms. Several run-time systems are provided by the Zerberus System, but to guarantee the generality of our approach the developer can also design his own run-time system, e.g. if the desired platform is not supported. To avoid a repeated implementation of such run-time systems, Zerberus offers a way to code such run-time systems application independent. By the use of other means, the Zerberus tags, locations in the code that have to be replaced with application dependent data can be marked. The replacement takes place in the code generating process.

To enable the simultaneous use of run-time systems implemented by the developer and of run-time systems provided by Zerberus in a N-Version programming system all protocols for the fault-tolerance mechanisms are provided. Thus the design effort for a new run-time system is also minimized.

3.5   Code Generation

The transformation of the functional model, the application dependent code and the selected fault-tolerance mechanisms into executable code is performed automatically by the Zerberus code generator. The whole code generation process is depicted in fig.2. Both the functional model and the run-time systems are parsed by the code generator and syntactic and semantic checks are performed. Afterwards the code generator replaces the Zerberus tags by application data and produces executable code.

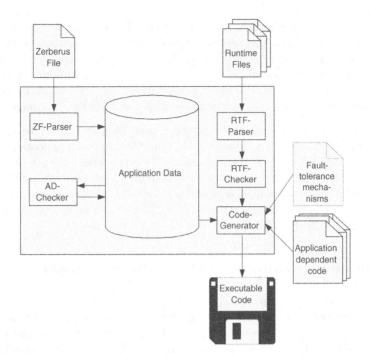

Fig. 2. Code generation process

3.6   Certification of the Zerberus System

The certification of an application developed with the Zerberus System can be
split up into three distinct parts:

1. Certification of the Zerberus System approach
2. Certification of the Zerberus run-time system
3. Certification of the application-dependent code

Certification of the Zerberus System Approach. In a first step the Zerberus System
tem approach has to be certified. This certification includes the Zerberus Systems
tems concept (including voting, synchronization, integration algorithms), the
Zerberus Language, the code generator and the Zerberus Tags. All tools are
currently available as prototypes. For a successful certification these tools have
to be re-engineered according to the standards proposed by the certification
authorities (RTCA,FDA,TÜV).

Certification of the Zerberus Run-Time Systems. In the second step the certification of the run-time systems is performed. This includes tests of the successful implementation of the proposed algorithms, the successful execution of
functional models and the conformance with the proposed standards of the certification authorities. Currently two prototype implementation for VxWorks 5.5
and the programming languages C and C++ are available.

Certification of the Application Dependent Code. For the certification of an application developed with the Zerberus System only a certification of the functional model, the code implemented by the user and the compliance with the Zerberus run-time system should be necessary. To achieve this minimization a strong partitioning among the different integrated modules must be ensured. This separation is another requirement towards the Zerberus Language.

For a successful certification the system must of course apply to the certification standards. These standards differ from the fields of application [22]. In general this means that the system must be re-engineered for each such standard. In case a certification is achieved the system can be reused for applications of the same domain without a repeated certification of the steps one and two. We intend to achieve such a certification by the German certification authority TÜV for the medical domain.

4   Informal Description of the Zerberus Language

In the previous section the requirements on the Zerberus Language were discussed in the context of the different development process steps. In this section the Zerberus Language is described informally and it is shown that the requirements can be satisfied by the Zerberus Language. The language was influenced by the language Giotto introduced in Berkeley [10]. Giotto was changed and extended in a way that the resulting Zerberus Language was suited for the use for fault-tolerant applications.

The main attribute to support voting, synchronization and integration algorithms is replica determinism. This is a non-trivial issue since different platforms can be used to achieve fault-tolerance. This includes the simultaneous use of different hardware, operating systems, programming languages and control algorithms in one control system. To achieve replica determinism nevertheless the Zerberus Language is based upon the time-triggered paradigm [5]. Similar to the approach in [23] replica determinism can be achieved by using the knowledge about the execution times. In the context of control applications the execution times can be related to the frequency of control cycles.

Basing the voting, synchronization and integration algorithms on the frequency of control cycles has different positive outcomes: by specified frequencies of control cycles in the functional model there exist on the one hand deterministic points in time, when the synchronization and voting algorithms can take place. On the other hand the execution and scheduling of the different processes can be carried out in different ways on the Zerberus units between these points.

The existence of deterministic points in time allow the application of distributed voting and synchronization algorithms. In this way a single point of failure can be avoided.

To achieve the claimed simplicity of the language, the Zerberus Language consists of only seven different object types: ports, actors, sensors, guards, modes and modechanges. In this section the different object types are explained informally.

4.1 Port

All communication in the Zerberus System is performed via ports. A port is a unique space in memory with a predetermined size and a specified representation. Port types are the only element of the Zerberus Language, that refer directly to a specific platform. To guarantee the platform independence the port types are platform independent, but are based on the fundamental types of the most common programming languages.

The values of the ports represent the state of the Zerberus units. Therefore a comparison of the different Zerberus units can be based on the values of these ports. It is required that there are no spaces in memory to store internal states besides the ports. Thus the state synchronization can also be based on the values of the ports during the reintegration of a Zerberus unit. The platform independent specification of the size and the representation of the port values is the foundation to enable the use of N-Version programming using different programming languages and operating systems.

In the following the attributes of ports are described. Ports are persistent, that means a port keeps its value over time until the port is updated. The update access has to be performed deterministically: it is not allowed that more than one write access is performed at a certain point in time. This condition is checked by the code generator while parsing the functional model and in addition at run-time (necessary due to the possible usage of guards, see section.4.6).

Replica non-determinism can also be the result of small clock differences (since the synchronization algorithm can only guarantee a deviation of the local clock from the global clock smaller than ε) or of N-Version programming. Due to these effects the correct port values are typically situated in a small interval. To support this fact the comparison of ports can also be based on an interval decision. This can be done by declaring a voting function for the port that has to be implemented by the developer. In case no voting function is specified the voting of the port values is based on the bit-by-bit comparison.

The voting on the value of a specific port takes place at least every time an output is performed based on this port value. For a faster detection of errors the developer can also specify shorter voting intervals.

4.2 Task

The separation of the pure functionality of the application and the run-time system including the fault-tolerance mechanisms is realized by tasks. Tasks are periodically called functions and realize the actual control system functionality. The simultaneous execution of different tasks is allowed, but to achieve determinism in the execution the tasks have to be independent of each other and synchronization points are not allowed. Thus the implementation of the task functions is simplified and accelerated since they represent only sequential programs and the requirement of the strict partitioning of the integrated modules to reduce the certification effort is satisfied.

The communication of the tasks between each other and with the environment is exclusively performed via ports. The access of tasks on ports occurs in a time-triggered manner. At the beginning of every invocation the task reads the values of the input ports, at the end of the invocation the results are written into the output ports of the task. Here the begin and the end refers to the invocation period as specified in the functional model. The port access is realized by the Zerberus run-time system and is performed in logical zero time.

The actual execution of the task on the CPU is scheduled by the Zerberus run-time system and is transparent to the developer. Nevertheless the developer has to guarantee that the worst-case execution times (WCETs) of the tasks allow a completion of the tasks satisfying the temporal restrictions as specified in the functional model.

4.3   Sensor and Actor

Sensors and actors realize the communication of the application with the environment and should not be mistaken for the hardware devices. Sensors are functions that are executed to read values from the environment and to write these values into ports, actors are functions to read values from the port and write these values to the environment.

The execution of the sensor and actor functions is also performed time-triggered. The execution frequency has to be specified by the developer. The sensor execution takes thereby place at the begin of each interval, the actor execution at the end of each interval. Both executions are regarded as instantaneous. To legitimate this assumptions the functions must represent short sequential code without synchronization points and blockages. For example in case of a network device the sensor functions may check the arrival of a message and copy the message into a port but a blockage until the receive event of a new message is not allowed.

4.4   Mode

Applications can have different operation modes. To support this feature the Zerberus Language introduces modes. A mode is a set of tasks, sensors and actors that is currently active on the Zerberus units. In addition, a mode cycle duration is assigned to every mode. Within each mode cycle the tasks, sensors and actors are executed according to their frequency as specified in the mode declaration.

```
mode m
{
        task= t1 1,t2 2;
        actor= a 2;
        sensor= s 1;
        duration= 50000000 ns;
}
```

Fig. 3. Mode declaration

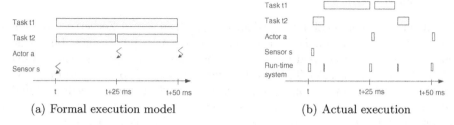

(a) Formal execution model   (b) Actual execution

Fig. 4. Execution model for mode m

Figure 3 shows the declaration of an example mode m in the Zerberus Language. m contains two tasks t1 with frequency 1 and t2 with frequency 2, a sensor s with frequency 1 and an actor with frequency 2. The duration of one mode cycle is set to 50 ms.

The formal execution model is depicted in figure 4(a) under the assumption that the mode cycle starts at time t. At time t the function of sensor s is executed and the tasks t1 and t2 are started. At time t+25ms the task t2 is stopped and the actor function is executed. Afterwards the task t2 is started for a second time. At the end of the mode cycle at t+50ms both tasks are stopped and the actor a is executed a second time. The execution of the sensor and actor functions appear instantaneous in the execution model.

Figure 4(b) shows a possible actual execution of the mode cycle on the machine. In addition to the task execution also the time required for the actor and sensor function execution, as well as the time consumed for run-time system execution have to be considered. The run-time system realizes the scheduling of the tasks, the port accesses and the voting and synchronization with the other Zerberus units.

The scheduler used in the example of fig. 4(b) uses a Earliest-Deadline-First strategy for the task execution. Sensors and actors are executed within the run-time system context.

4.5 Modechange

To enable the switch between different operation modes modechanges can be used. A modechange is a function implemented by the developer that evaluates if a mode should be switched or not. The developer has to specify the target mode and a non-empty set of source modes within the modechange declaration. The evaluation of the function, which is based on the values of the assigned ports, takes always place at the end of the source mode cycles.

Mode switches must be deterministic, this means that for every achievable configuration (port values and modes) at most one assigned modechange can reach a positive evaluation for a modechange. This condition is checked in the Zerberus System at run-time.

4.6 Guard

Guards are another possibility to change the behavior of a Zerberus program. Guards are similar to modechanges functions based on port values, but while modechanges should be used for different operation modes, guards can be used to control individual tasks. Thereto the guard is assigned to a certain tasks. At the begin of every invocation of this task, the guard function is evaluated and only in case of a positive evaluation the according task is started. The main advantage of guards over modechanges is therefore their flexibility. A guard can be used also within a mode cycle and not only at the end of the mode cycle.

5   Formal Description of the Zerberus Language

The concrete language specification is given in [24]. In this chapter we describe the language in a more abstract way. A Zerberus program computes on the base of some inputs by the environment the output to the environment. In the following we refer to $Input$ for the values of the environment inputs and $Output$ for the values of the output to the environment.

A Zerberus program consists of:

1. A set of port declarations: A port declaration $(p, type, init, comp)$ consists of a port name p, a type $type$, an initial value $init \in type$ and a compare mode $comp$. The set of allowed types are the basic types of common programming languages (abstracted to achieve platform independence) and arrays of fixed size of these types. Every port declaration must also contain an initial value to achieve a common start configuration for all units.

   The developer can specify how a port is treated within the voting algorithm. These possibilities range from the denial of comparisons, a bit-by-bit comparison to an user-defined comparison (typically an interval test). The denial of comparisons is only valid if the port is not read by an actor.

   Port names must be uniquely declared: that means if $(p, *, *, *)$ and $(p', *, *, *)$ are distinct port declarations, then $p \neq p'$.

   We refer to the set of declared ports by $Ports$, to the initial value of a port p by $init[p]$ and to the values of a set of ports $P \subseteq Ports$ by $Vals[P]$.

2. A set of actor declarations: An actor declaration (a, f, P) consists of an actor name a, an actor function name f and a set $P \subseteq Ports$ of input ports. Actor names must be uniquely declared: that means if $(a, *, *)$ and $(a', *, *, *)$ are distinct port declarations, then $a \neq a'$.

   The developer has to implement an actor function with the name f for each platform used. The function must be of the form $f : Vals[P] \rightarrow Output$ and is executed every time the actor is invoked synchronously within the system's context. We write $Actors$ for the set of declared actors and f_a for the function of an actor a.

3. A set of sensor declarations: A sensor declaration (s, f, P) consists of a sensor name s, a sensor function name f and a set $P \subseteq Ports$ of output ports. Sensor names must be uniquely declared.

The developer has to implement a sensor function with the name f for each platform used. This function must be of the form $f : Input \rightarrow Vals[P]$. The sensor function is executed every time the sensor is invoked synchronously within the system's context. We refer to $Sensors$ for the set of declared sensors, to f_s for the function of a sensor s and to $res_s[p]$ for the results regarding port $p \in P$ of the sensor function.

4. A set of guard declarations: A guard declaration (g, f, P) consists of a guard name g, a guard function name f and a set $P \subseteq Ports$ of evaluation ports. Guard names must be uniquely declared.

The developer has to implement a guard function with the name f for each platform used. This function must be of the form $f : Vals[P] \rightarrow \mathbb{B}$. Guard functions are invoked every time the assigned task should be started. The execution of the guard function takes place synchronously within the systems context. We write $Guards$ for the set of declared guards, f_g for the function of a guard g, $p[g]$ for P and $res_g(Vals[P])$ for the results of one function invocation based on the current values of the assigned ports.

5. A set of task declarations: A task declaration $(t, f, g, In, Out, Inout)$ consists of the task name t, the task function name f, optionally a guard $g \in Guards$ and a set of Ports $In \cup Out \cup Inout \subseteq Ports$. Task names must be uniquely declared.

The set of assigned ports is subdivided into three classes: In, Out and $Inout$. These classes refer to the access type of the task to the port. Every port used in the task must belong to exactly one class.

The developer has to implement the task function with the name f for each platform used. The function must be of the form $f : Vals[In \cup Inout] \rightarrow Vals[Out \cup Inout]$ and is performed every time the task is invoked by the system. The execution takes place asynchronously to the system's context.

We write $Tasks$ for the set of declared tasks, $res_t[p]$ for the results of the current function invocation of task t concerning one assigned output port $p \in In \cup Out$ and f_t for the function of a task t.

6. A set of mode declarations: A mode declaration $(m, start, T, A, S, d)$ consists of a mode name m, a boolean value $start$, task assignments T, actor assignments A, sensor assignments S and a duration d. Mode names must be uniquely declared.

Within the application exactly one mode must be declared as start mode m_{start}, that means $start = true$. The system will start the operation in this mode.

A task assignments $(t, freq)$ consists of a task $t \in Tasks$ and a related frequency $freq \in \mathbb{N}$. The frequency determines the number of the task invocations within one mode cycle (except if a related guard evaluates false). In the following we will refer to the frequency $freq$ of a task t in mode m by $freq(t, m)$. The sensor and actor assignments are similar.

The duration (s, ns) consists of the number of seconds $s \in \mathbb{N}$ and the number of nanoseconds $ns \in \mathbb{N}$ (to confirm with the POSIX standard) and determines the duration of one mode cycle.

We write $Modes$ for the set of declared modes.

7. A set of modechange declarations: A modechange declaration $(mc, f, P, Source, target)$ consists of the modechange name mc, a modechange function name f, a set $P \subseteq Ports$ of evaluation ports, a set of source modes $Source \subseteq Modes$ and a target mode $target \in Modes$. Modechange names must be uniquely declared.

The developer has to implement a modechange function with the name f for each platform used. The function must be of the form $f : Vals[P] \to \mathbb{B}$. A modechange is evaluated always at the end of a mode $m \in Source$. If the function result is true the new mode executed by the system will be $target$. We write $Modechanges$ for the set of declared modechanges, f_{mc} for the function of a modechange mc, $p[mc]$ for P and $res_{mc}(Vals[P])$ for the results of one function invocation.

In the following the semantics of the Zerberus Language are described. The realization of the fault-tolerance mechanisms is mentioned but the focus lies on the functional semantics.

The voting algorithm has three results: the state of the system $res_{sys} \in \mathbb{B}$, the state of the own unit $res_{unit} \in \mathbb{B}$ and the acting unit id $act \in \mathbb{N}$. The result of the synchronization is the temporal correction value Δ_{cor}. In addition we assume that the developer has decided to use the port values for voting only in case they are used for an actor output.

For simplicity reasons possible occurrences of errors during the application execution are ignored. These errors can be time violations or simultaneous write attempts on one port. In all such cases the normal execution is aborted at once and fault reaction algorithms are executed.

A program configuration $C = (id, s\_sys, s\_unit, m, \delta, v, \sigma_{active,}, \tau)$ consists of the unique Zerberus unit ID id, states of the system s_sys and of the own unit s_unit, a current mode $m \in Modes$, a mode unit $\delta \in \mathbb{N}$, a valuation $v \in Vals[Ports]$ for all ports, a set of active tasks $\sigma_{active} \subseteq Tasks$ and a time stamp $\tau \in \mathbb{Q}$. The set of active tasks σ_{active} contains all tasks that are logically running, whether or not they are physically running by expending CPU time. The mode unit δ represents the current internal point of the mode cycle. The number of internal points within one mode cycle of mode m is determined by the least common multiple $\omega[m]$ of the frequencies of the tasks, actors and sensors assigned to m.

At start-up each Zerberus unit has to determine if the system is currently running or if an initial synchronization procedure must be started. This is realized by a function of the run-time system that observes the network. An operating system can be recognized by voting and synchronization messages.

In case the system is already running another run-time system function is executed that allows to obtain the states of the other Zerberus units. One requirement for a state synchronization is that the system is currently at the beginning of one mode cycle ($\delta = 0$). In this case no tasks are active on the other units and an integration can be successful. Another requirement is that the majority of Zerberus units agrees in their states. If both requirements are met the configuration is updated to the state of the majority and the integration was successful.

If on the other hand the system is not running an initial synchronization procedure is started. The goal of this procedure is to obtain a global time base. In case of a successful synchronization the initial configuration is set to $C_{init} = (id, true, true, m_{start}, 0, v_{init}, \emptyset, \tau_i)$ where τ_i is the result of the initial synchronization and $v_{init}[p] = init[p]$.

The internal points represent the points in time, when the synchronization and voting algorithms are executed. At each internal point the following steps are performed by the run-time system on the basis of the current configuration $C = (id, s\_sys, s\_unit, m, \delta, v, \sigma_{active}, \tau)$:

1. Copying of task results: Let $\sigma_{completed}$ be the set of tasks $t \in Tasks$ that are completed. A task t is completed if $t \in \sigma_{active}$ and if δ is an integer multiple of $\omega[m]/freq(t,m)$ at configuration C. For all ports $p \in Ports$: if $p \in inout[t] \cup out[t]$ of a task $t \in \sigma_{completed}$ then define $v_{stop}[p] = res_t[p]$, else $v_{stop}[p] = v[p]$. Let C_{stop} be the new configuration that agrees with v_{stop} in the values of ports and with the set of active tasks $\sigma_{stop} = \sigma_{active} \setminus \sigma_{completed}$ and otherwise agrees with C.

2. Voting and synchronization: Let $a_{execute}$ be the set of actors to be executed. An actor a is executed if δ is an integer multiple of $\omega[m]/freq(a,m)$ at configuration C_{stop}. Let p_{vote} be the set of all ports read by the actors $a \in a_{execute}$. The voting and synchronization algorithms are then invoked with the parameters $v_{stop}[p_{vote}]$, the mode m and the mode unit δ. Let res_{system}, res_{unit} and act be the results of the voting algorithms and Δ_{cor} be the result of the synchronization algorithm. If $((res_{system} \wedge res_{unit}) = false) \vee (|\Delta_{cor}| > \epsilon)$ then the normal system execution is aborted and error reaction and recovery algorithms are invoked. Otherwise let C_{vote} be the new configuration that agrees with $s\_sys_{vote} = res_{system}$, $s\_unit_{vote} = res_{unit}$, $act_{vote} = res_{act}$ and $\tau_{cor} = \tau + \Delta_{cor}$ and otherwise agrees with C_{stop}.

3. Execution of actors: Let $a_{execute}$ be the set of actors to be executed. For all actors $a \in a_{execute}$ the actor function f_a is executed if $id = act_{vote}$. If $id \neq act_{vote}$ the unit only controls the correct output (performed by another unit). In case errors are detected by the system error recovery algorithms are executed. The execution of the actor functions takes places synchronously within the run-time system execution that means that the run-time system waits for the completion of the actor function. Let C_{actor} be the new configuration that agrees with C_{vote}.

4. Evaluation of modechanges: If $\delta = 0$ modechanges have to be evaluated. The set of modechanges mc_{eval} that needs to be considered consists of all modechanges mc with $m \in source(mc)$. For each modechange $mc \in mc_{eval}$ the corresponding function is evaluated and if $f_{mc}(v_{stop}[p[mc]]) = true$ then $m' = target(mc)$. The developer has to guarantee that at most one modechange evaluates true at a time. The run-time systems checks this condition and creates an internal error in case of a violation of this rule. In the latter case the system execution is stopped and fault reactions are started.

If no modechange evaluates true, then $m' = m$. Let $C_{modechange}$ be the new configuration that agrees in m' as new operating mode and otherwise with C_{actor}.

5. Execution of sensors: Let $s_{execute}$ be the set of sensors s to be executed. A sensor is executed if δ is an integer multiple of $\omega[m']/freq[s, m']$. Let p_{sensor} be the set of ports that are written by a sensor $s \in s_{execute}$. For each port $p \in Ports$: if $p \in p_{sensor}$ $v_{sensor}[p] = res_s[p]$, else $v_{sensor}[p] = v_{stop}[p]$. Let C_{sensor} be the new configuration that agrees with v_{sensor} in the values of the ports and otherwise with $C_{modechange}$.

6. Invocation of tasks: Let t_{start} be the set of tasks t to be started. A task t is started if δ is an integer multiple of $\omega[m]/freq[t, m']$. In addition if the task has a guard the evaluation must be positive: $res_g(v_{sensor}[p[g]]) = true$. For every task $t \in t_{start}$ the function f_t is invoked with the specified parameters based on the values v_{sensor}. Let C_{start} be the new configuration that agrees with the set of active tasks $\sigma_{start} = \sigma_{stop} \cup t_{start}$ and otherwise with C_{sensor}.

7. Advance time: If $\delta = \omega[m] - 1$ then $\delta' = 0$ otherwise $\delta' = \delta + 1$. Let $\tau' = \tau_{cor} + d[m']/\omega[m]$. The next time the program is invoked with step 1 is at time τ'. Let C_{succ} be the new configuration that agrees with δ' and τ' and otherwise with C_{start}.

6   Case Study

For demonstration we have implemented a system to balance a rod under the control of switched solenoids, see figure 5. For a stable control sample rates in the range of few milliseconds are necessary. As device an AD/DA-board was used to connect the experimental setup with the three computer units. The computers were equipped with AMD Athlon processors and they were connected by switched ethernet. As real-time operating system we used VxWorks and as programming language C.

Fig. 5. Balanced rod

```
/* Code for the rod control*/

/*ports*/
port input
{
        type=INT16;
        compareTIME=NEVER;
        initialValue=0;
}

port param
{
        type=INT16[2];
        compareTIME=NEVER;
        initialValue=0;
}

port output
{
        type=INT16;
        compareTIME=compare();
        initialValue=0;
}
```

```
/*actors and sensors*/
sensor sens
{
        function=read();
        out=input;
}

actor act
{
        function=write();
        in=output;
}

/*tasks*/
task control
{
        function: contron();
        in= input;
        inout=param;
        out=output;
}

mode control_cycle
{
        startmode;
        task: control 1;
        sensor: sens 1;
        actor: act 1;
        duration: 1000000 ns;
}
```

Fig. 6. Functional model

The implementation of the control program was done by two students. It took two weeks to implement the PID controller on a single computer. The conversion of the code to the Zerberus System and the addition of the fault-tolerance mechanisms could be realized within two hours using the code for the single-machine version. The code that had to be implemented for the fault-tolerant controller was less than 100 lines of code.

For describing the functional model of the control application 30 lines of code in the Zerberus Language were needed. The code is depicted in figure 6. Three ports had to be declared: one port for the systems input (the deviation of the current position from the desired position), an array of two integer values for the differential and integral part and one port for the result. Only the port for the result was used for the voting algorithm. Also the rest of the functional model was very simple: a sensor was used to read the current position of the rod, a task was needed for the control computation and an actor was used for writing the output to the environment.

In addition to the functional model four functions were needed for the control program:

- *read()*: The sensor function was used to read the current value from the AD/DA-board.
- *control()*: This function implemented the PID controller. As input the function uses the current position of the rod. The function computes the necessary control output for stabilizing the rod at the desired position. To achieve this goal the function uses two further ports to obtain also the differential part and the integral part of the controller.
- *compare()*: The function *compare()* is used within the execution of the voting algorithm of the run-time system. Due to synchronization differences and to sensor imprecision a binary compare of the result of the *control()* function was not possible. Therefore the two students implemented an interval decisions: two results were assumed to be correct if the difference between both values was less than 0.1 V (allowed voltage range was -10..10 V).
- *write()*: The actor function was used to write the value of the port output to the AD/DA-board.

The code for these functions consisted of less that 70 lines of code.

The addition of the fault-tolerance mechanisms (voting, synchronization, integration), the communication between tasks, sensors and actors, as well as the scheduling was realized by the system. The sample rate for this control example was 1000 Hz.

This example proves the applicableness for small control applications. However we are currently working on two pilot projects with the industry. The goals of these projects are on the one hand to point out the feasibility, but on the other hand also to adopt industrial standards in the Zerberus System to increase the acceptance rate in the industry.

7   Conclusions and Future Research

In this paper we have introduced the Zerberus Language. This language enables the developer to design the functional model of the control application. The design of the language was guided by the different requirements on the language and the development system.

To achieve a general applicability the constraints by the language should be minimized. This was realized by the independency from a certain platform and by the time-triggered approach which is suitable for most control systems.

For the use with fault-tolerance mechanisms and especially with active redundancy the language must provide features for replica determinism. By the time-triggered approach this requirement is satisfied. In addition deterministic points in time for the execution of voting algorithms are available and also a synchronization of the different units can be achieved. The state synchronization during the reintegration phase is enabled by separating the inner state (ports) from the functionality (tasks).

One main aspect of supporting the acceleration of the certification process is the strict separation of the different integrated modules. This separation is realized by the task concept of the Zerberus Language. In case operating systems are used that support memory protection, it can be guaranteed that the run-time system is not influenced by the tasks except in the predefined way.

The Zerberus Language is therefore suited for the use within the Zerberus System. A code generator is available to support the transformation of the functional models designed in the Zerberus Language into executable code. Within one small case study we demonstrated the usage of the Zerberus System.

To point out the applicableness within industrial projects we are currently working on two pilot projects with the industry. Within these projects we also plan to adopt the recommended development process and the tools to industrial standards. In addition we want to support further fault-tolerance mechanisms despite active structural redundancy. Therefore we intend to introduce another language to specify points within the execution when fault-tolerance mechanisms should be executed (events) and exception handlers to address the occurrence of failures. The goal is to provide a set of standard fault-tolerance mechanisms to the user. To assist the developer in choosing adequate mechanisms, guidelines will be developed.

Another research area will be an advanced support of the user in the certification process. Document output automated by the used tools and the compliance of tools and run-time systems with the relevant development standards are planned. Within one project for a medical control system in cooperation with the German certification authority TÜV we want to exemplify our approach.

References

1. Pradhan, D.K.: Fault-Tolerant Computer System Design. Prentice Hall (1996)
2. Lee, P.A., Anderson, T.: Fault Tolerance: Principles and Practice. Springer-Verlag New York, Inc., Secaucus, NJ, USA (1990)
3. RTCA DO-178B: Software considerations in airborne systems and equipment certification (1992)
4. International Electrotechnical Commission: IEC 61508: Functional safety of electrical/electronic/programmable electronic safety-related systems. (1998)
5. Kopetz, H., Bauer, G.: The Time-Triggered Architecture. Proceedings of the IEEE **91** (2003) 112 – 126
6. TTTech Computertechnik AG: Time Triggered Protocol TTP/C High-Level Specification Document. (2003)
7. Kopetz, H., G.Grnsteidl, J.Reisinger: Fault-tolerant membership service in a synchronous distributed real-time system. In: Dependable Computing for Critical Applications. (1991) 411–429
8. Kopetz, H., Fohler, G., Grünsteidl, G., Kantz, H., Pospischil, G., Puschner, P., Reisinger, J., Schlatterbeck, R., Schütz, W., Vrchoticky, A., Zainlinger, R.: The distributed, fault-tolerant real-time operating system mars. IEEE Operating Systems Newsletter **6** (1992)

9. Website DECOS: ⟨http://www.decos.at/⟩
10. Henzinger, T.A., Horowitz, B., Kirsch, C.M.: Giotto: A time-triggered language for embedded programming. Proceedings of the First International Workshop on Embedded Software (EMSOFT) (2001) 166 – 184
11. Henzinger, T.A., Horowitz, B., Kirsch, C.M.: Embedded control systems development with giotto. Proceedings of the International Conference on Languages, Compilers, and Tools for Embedded Systems (LCTES) (2001) 64 – 72
12. Berry, G., Gonthier, G.: The esterel synchronous programming language: Design, semantics, implementation. Science of Computer Programming 19 (1992) 87–152
13. Caspi, P., Pilaud, D., Halbwachs, N., Plaice, J.A.: Lustre: a declarative language for real-time programming. In: POPL '87: Proceedings of the 14th ACM SIGACT-SIGPLAN symposium on Principles of programming languages, ACM Press (1987) 178–188
14. Amnell, T., Fersman, E., Mokrushin, L., Pettersson, P., Yi, W.: Times - A Tool for Modelling and Implementation of Embedded Systems. In: Joint European Conferences on Theory and Practice of Software, ETAPS 2002. Lecture Notes in Computer Science, Springer-Verlag (2002)
15. Krcal, P., Yi, W.: Decidable and Undecidable Problems in Schedulability Analysis Using Timed Automata. In: Joint European Conferences on Theory and Practice of Software, ETAPS 2004. Lecture Notes in Computer Science, Springer-Verlag (2004)
16. Armstrong, J.: Erlang — a Survey of the Language and its Industrial Applications. In: INAP'96 — The 9th Exhibitions and Symposium on Industrial Applications of Prolog, Hino, Tokyo, Japan (1996) 16–18
17. Armstrong, J.: The development of erlang. In: ICFP '97: Proceedings of the second ACM SIGPLAN international conference on Functional programming, New York, NY, USA, ACM Press (1997) 196–203
18. Echtle, K.: Fehlertoleranzverfahren. Springer Verlag (1990)
19. Lamport, L., Melliar-Smith, P.M.: Synchronizing clocks in the presence of faults. J. ACM 32 (1985) 52–78
20. Lundelius, J., Lynch, N.A.: A new fault-tolerant algorithm for clock synchronization. In: Symposium on Principles of Distributed Computing. (1984) 75–88
21. Schmid, U., Schossmaier, K.: Interval-based clock synchronization. Real-Time Systems 12 (1997) 173–228
22. Saglietti, F.: Licensing reliable embedded software for safety-critical applications. Real-Time Systems 28 (2004) 217–236
23. Poledna, S., Burns, A., Wellings, A., Barrett, P.: Replica determinism and flexible scheduling in hard real-time dependable systems. IEEE Transactions on Computers 49 (2000) 100–110
24. Buckl, C.: Zerberus Language Specification Version 1.0. Technical Report TUM-I0501, TU München (2005)

Soft Error Mitigation in Cache Memories of Embedded Systems by Means of a Protected Scheme

Hamid R. Zarandi and Seyed Ghassem Miremadi

Department of Computer Engineering, Sharif University of Technology
zarandi@ce.sharif.edu, miremadi@sharif.edu

Abstract. The size and speed of SRAM caches of embedded systems are increasing in response to demands for higher performance. However, the SRAM caches are vulnerable to soft errors originated from energetic nuclear particles or electrical sources. This paper proposes a new protected cache scheme, which provides high performance as well as high fault detection coverage. In this scheme, the cache space is divided into sets of different sizes. Here, the length of tag fields associated to each set is unique and is different from the other sets. The other remained bits of tags are used for protecting the tag using a fault detection scheme e.g., generalized parity. This leads to protect the cache without compromising performance and area with respect to the similar one, fully associative cache. The results obtained from simulating some standard trace files reveal that the proposed scheme exhibits a performance near to fully associative cache but achieves a considerable fault detection coverage which is suitable to be used in the dependable computing.

1 Introduction

Designer of embedded microprocessors have to compromise between performance, cost and energy. Memory hierarchy is one of the most important elements in modern embedded systems. In particular, cache memories are simple cost-effective elements to achieve higher memory bandwidth, which significantly affects the peak throughput [12]. The performance of the cache depends on several factors such as cache size, block size, mapping function, replacement algorithm, and write policy [18]. Recent modern processor designs often devote a large fraction of on-chip transistors (up to 80%) to caches [15]. Consequently, the reliability of caches affects the dependability of the overall system. The purposes of integrating an error checking scheme in the memory system such as caches are to prevent errors to propagate to other components and to overcome the effects of errors locally. This contributes the overall goal of achieving failure-free computation.

There are two important error checking schemes: parity codes [22] and error-correcting codes (ECC) [9]. Parity codes can detect odd number of errors and their power consumption is much less than ECC [4], however, ECC can correct errors. Each of these codes has a serious problem, i.e., the parity codes have low error detection capability [25], and the ECC codes incur low performance [17]. Moreover, using each of them in a uniform structure may occupy significant area space [7], [11]. It is

C.A. Maziero et al. (Eds.): LADC 2005, LNCS 3747, pp. 121 – 130, 2005.

due not to flexible in terms of chip area requirements as the area occupied by them is directly proportional to the cache size. For example, 12.5% area overhead is needed to store a parity for 8-bit data and the same overhead is required for an 8-bit SEC-DED (single error correction, double error detection) for a 64-bit entity. It should be noted that these codes could detect or correct only single error and have not good efficiency in the applications that are more prone to multiple-errors e.g., space applications.

This paper introduces a new placement scheme for cache memories based on a variable associativity degree. This scheme is a generalized version of the HBAM cache (Hierarchical Binary Associative Mapping), which is previously introduced in [23], [35]. In this scheme, using a division parameter k, cache space is divided into sets of different sizes, similar to set-associative one, but organized in a hierarchical structure where the size of the set at a given level is k times larger than that of the set in the next level of hierarchy. Thus, this new scheme is called *Hierarchical Multiple Associative Mapping* (*HMAM*). Unlike set-associative mapping with a fixed modulo-based address translation from CPU address into cache sets, HMAM uses an address translation function that behaves based on a variable modulo system. This characteristic leads to increase hit ratio and to decrease both area and power consumption regarding to the fully-associative caches. Using this algorithm, a new solution to the mentioned problem can be obtained since it enables us to provide a protection code such as generalized parity for every tag of cache line to detect multiple faults without compromising the performance or increasing significant area.

The remained of the paper is organized as follows. Section 2 presents some related work. Section 3 gives overview of the problem, the proposed cache architecture and fault detection method. Section 4 experimentally studies performance and fault detection of our method. Section 5 discusses hardware complexity. Finally, section 6 concludes the paper.

2 Related Work

A uniprocessor may have a large miss penalty when it has only a first-level cache and the gap between processor and memory speed is large. Increasing associativity also has the advantage of reducing the probability of thrashing. Repeatedly accessing m different blocks that map into the same set will cause thrashing. A cache with an associativity degree of n can avoid such thrashing if $n \geq m$ and LRU replacement policy is employed [27]. A hash-rehash cache [1] uses two mapping functions to determine the candidate location with associativity of two and by sequential search, but higher miss rate results from non-LRU replacement.

Agarwal et al. [2] proposed the column-associative cache that improves the hash-rehash cash by adding hardware to implement LRU replacement policy. The predicative sequential associative cache proposed by Calder et al. [6], uses bit selection, a sequential search and steering bit table, which is indexed by predictive sources to determine search order. However this approach is based on prediction, which may be incorrect and has slightly longer average access time. Skewed-associative cache [16] increases associativity in orthogonal dimension using skewing function instead of bit selection to determine candidate locations. The major drawbacks of this scheme are a longer cycle time and the mapping hardware necessary for skewing. Ranagathan [15] proposed a configurable cache architecture useful for media processing which divide the cache into partitions at the granularity of the conventional cache. The key drawback of it is that the number and granularity of the partitions are limited by the associativity of the cache and also it causes to modify the hardware of the cache to support dynamic partitioning and associativity.

Another configurable cache architecture has been proposed in [26], which intended for some specific applications of embedded systems. The cache mapping function can be configured by software to be direct mapped, 2-way, or 4-way set-associative caches.

Also, several studies have been done to provide fault-tolerance in caches memories [11], [17], [25]. In [11], a very small cache was proposed to store parity information or to duplicate recently used data with a very good hit rate. In [17], a programmable address decoder was proposed to disable faulty blocks and to remap their references to non-faulty blocks. But area overhead for a typical 16KB cache is 11% of total cache area. Replicating data in the cache to enhance reliability was proposed in [25]. The fault detection scheme was either parity or ECC, and in the case of detecting faults, one of the replications of the affected word in the cache was used. However, this scheme can detect only single transient faults and has significant effects on the performance of the cache such that miss rate of the cache increases up to 4 times.

3   Problem Overview

Use of lower voltage levels, high speed data read/write operations and extremely dense circuitry increase the probability of transient fault occurrence, resulting in more bit errors in cache memories. Moreover, external disturbances such as noise, power jitter, local heat densities, and ionization due to alpha-particle hits can also corrupt the information [10], [13].

Most of transistors in a cache are in memory cells. Hence, the probability that a given defect is in a memory cell (a bit in the data or tag field) is higher than the probability of it being in the logic circuitry.

Furthermore, the faults occurred in a tag field are more serious than those affect on the data field. It is due to 1) the size of a CAM cell is about double as that of a RAM cell [14], 2) each tag entry is responsible for storing/retrieving several words in its corresponded block. This means that a given fault in the tag has B (block size) times crucial effects more than the similar one occurred in a word of the corresponded block, and 3) bit changes in the tag cause the improper cache hit and miss decisions i.e., pseudo-hit, multi-hit, and pseudo-miss, and make the memory references to be invalid. In the case of a pseudo-hit, the processor gets wrong data on a read and updates the data in the wrong location on a write. A pseudo-miss generates an unnecessary main memory access. The multi-hit may be detected by the cache controller but handling is not simple. The controller cannot distinguish between the multiple hit lines to service the processor's request.

Parity codes are extensively used in cache memories of today's modern processors. As an example, the parity checking employed in data cache in the Pentium® processors [20] are: 1) parity bit per byte in the cache storage RAM, and 2) parity per entry in the tag array. However, parity can only detect odd number of errors (coverage of 50%) and is not suitable for the applications which need high reliability. For example, a relatively large fraction of the transient faults caused by alpha-particle radiation or heavy-ion [10] manifests as multiple-bit errors i.e., single-event multiple upsets [3] [21].

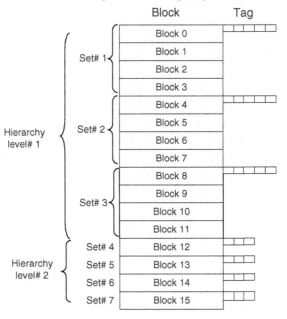

Fig. 1. The 4HMAM organization of a 16-block cache in a system with a 128-block main memory

Table 1. Address mapping and required bits for tag in each set

Set #	Logical condition in address decoder	Tag storage bits	Bit length of tag array	set Asso.
1	$A_{b+1}A_b = 1$	$A_{n-1} downto A_{b+2}$	$n - b - 2$	$C/4$
2	$A_{b+1}\overline{A_b} = 1$	$A_{n-1} downto A_{b+2}$	$n - b - 2$	$C/4$
3	$\overline{A_{b+1}}A_b = 1$	$A_{n-1} downto A_{b+2}$	$n - b - 2$	$C/4$
4	$A_{b+3}A_{b+2}\overline{A_{b+1}}\,\overline{A_b} = 1$	$A_{n-1} downto A_{b+4}$	$n - b - 4$	$C/4$
...	...	...	...	...
$3\log_4 C - 3$	$\overline{A_{b+2\log_4^C - 3}}A_{b+2\log_4^C - 4}...\overline{A_b} = 1$	$A_{n-1} downto A_{b+2\log_4^C - 2}$	$n - b - 2(\log_4 C - 1)$	4
$3\log_4 C - 2$	$A_{b+2\log_4^C - 1}A_{b+2\log_4^C - 2}...\overline{A_b} = 1$	$A_{n-1} downto A_{b+2\log_4^C}$	$n - b - 2\log_4 C$	1
$3\log_4 C - 1$	$A_{b+2\log_4^C - 1}\overline{A_{b+2\log_4^C - 2}}...\overline{A_b} = 1$	$A_{n-1} downto A_{b+2\log_4^C}$	$n - b - 2\log_4 C$	1
$3\log_4 C$	$\overline{A_{b+2\log_4^C - 1}}A_{b+2\log_4^C - 2}...\overline{A_b} = 1$	$A_{n-1} downto A_{b+2\log_4^C}$	$n - b - 2\log_4 C$	1
$3\log_4 C + 1$	$\overline{A_{b+2\log_4^C - 1}}\,\overline{A_{b+2\log_4^C - 2}}...\overline{A_b} = 1$	$A_{n-1} downto A_{b+2\log_4^C}$	$n - b - 2\log_4 C$	1

3.1 HMAM Organization

In the HMAM organization the cache space is divided into several variable size associative sets in a hierarchical form. Let C be the cache size in blocks and k be division factor used for dividing the cache space. In HMAM, the cache space is divided into k different sets with numbers of 1, 2, ..., k-1, k, with associativity of C/k located in hierarchy level of 1. However the last set i.e., k-th set, is then divided to k different associative sets in hierarchical level of 2 and with number of k, $k+1$, ..., $2k$-1. Nowthe last set i.e., $2k$-1 is then divided to k associative sets. This procedure is performed until the divided sets consist of only one block each. Hence, in HMAM cache, the cache size C should be power of k.

In this scheme, size of sets varies in power of k and number of sets is $(k-1)\log_k^C + 1$. Each set in this scheme with a hierarchy level of h has associativity of C/k^h blocks. The first k-1 sets have the largest size of C/k blocks while the last k sets contain a minimum of 1 block.

In this scheme, k is a parameter to adjust the associativity used in the cache. In other words, if k is 1 then the HMAM cache is a fully-associative cache and if k is C then the HMAM cache is a direct-mapped cache. Also, the HBAM cache [23] is a HMAM cache with k of 2. For typical size of k (i.e., 2, 4, 8, 16), the cost of this scheme is less than the cost of fully-associative scheme. Due to the separated logic associated to each set, its operation is relatively faster compared to fully-associative mapping and slower than set-associative mapping [33, 34]. In this scheme, address translation is performed dynamically (value based) instead of statically (bit selection) in the direct   mapping scheme. This means that there is no predefined specified format to determine set number of a memory address in the cache. The set number should be determined using the address pattern coming from the CPU. As an example, Fig. 1 shows the set organization of a 16-block HMAM cache with k equal to 4 (4HMAM), for a main memory of 128 blocks. Table 1 portrays the address mapping and required bits for tag storage in the 4HMAM cache where its block size is 2^b words. The number of sets in 4HMAM cache is $3\log_4^C + 1$.

3.2 Proposed Protection Code

In the HMAM cache, the reduced area, related to tag storage, was used for a protection code. The utilized protection code is named generic parity (GParity). A generalized parity in radix r, which also includes the even-parity when the radix is two, is sum of 1's in the word modulo r. It behaves as a checksum for the word and can detect any simultaneously single-event faults where number of faults is not divisible by r. As an example, the following figure show a 4HMAM tag of a memory address whose 4 bits are not necessary for storing. Its GParity in radix 16, which used for protecting this tag has been shown, as well. Its GParity is 5 since the number of 1's in the memory address is 5.

memory address	GParity	tag in the 4HMAM
0100 1001 1000 0100	0101	0 100 1001 1000

Fig. 2. A generalized parity in radix 8 for a memory address used in the 4HMAM cache

This protection code is more suitable for the HMAM because: 1) it can be easily adjusted to any length of bits, 2) it only depends on the number of 1's in the tag (or memory address) and hence can be calculated in parallel with address decoding, and 3) it can detect more portion of faults due to multiple effects of transient faults are all in the same event e.g., single-event multiple upsets [21].

In kHMAM cache, all tags of a set whose number is i, has $\log_2^k . \lfloor \log_k^i \rfloor$ bits length GParity. The cost of adder needed for calculating sum of 1's is negligible in total cost of the cache [8]. For example, a 32-bit adder synthesized via a common synthesis tool, use only 32 LUTs. Using the GParity in the HMAM cause the fault detection coverage to be improved and makes it to be suitable for dependable computing systems such as dependable embedded systems which have serious limitations in the area space.

4 Experimental Study

The cache simulator in [5] was modified to simulate the proposed HMAM cache scheme. Benchmarks used in this trace-driven simulation included several different kinds of programs of SPEC2000 benchmarks [19], namely *bzip2, apsi, swim, vortex, eon_cook, eon_rush, gcc, gzip, parser, sixtrack,* and *vpr.* Each file contains at least 100000000 references. Both data and instruction references are collected and used for the simulation. Three well-known placement schemes, i.e., the direct, set-associative, and fully-associative mapping are chosen for performance comparison and evaluation.

The cache miss-ratios for the conventional fully-associative (FA), 4-way set associative (4WSA), direct-mapped (DC) and the proposed HMAM cache with several values of k are shown in Fig. 3. For the fully-associative cache denoted as FA in the figure, the notation "32k-8byte" denotes an 32KB full-associative cache with a block size of 8 bytes. Notice the average miss ratio of the HMAM cache for a given size (i.e., 32KB) is very close to the FA. The HMAM cache is approaching to 4WSA when k is grown. The HMAM scheme outperforms the set-associative and direct mapping schemes for a wide variety of cache configurations.

4.1 Fault Injection and Fault Detection Coverage

For evaluation of the proposed protection code, we have simulated the scheme using the benchmarks in the presence of randomly injected faults and fault detection coverage of the protection code has been calculated. The fault model which has been used in the experiments was single bit-flip and multiple bit-flips. The number of bits affected had a uniform distribution. A typical cache size 32Kbyte with block size of 8 byte has been used. In this evaluation six cache schemes, 4WSA, 16HMAM, 8HMAM, 4HMAM, 2HMAM, and FA were equipped by GParity with the same hardware cost, and their performance and coverage were evaluated.

Table 2 portrays the number of bits and radix of GParity used in these caches and also shows their experimentally calculated fault coverage. All the single bit errors are detected by the protected code in every considered cache. The coverage of GParity when the radix is high (in set-associative), is near to 99% while for the 2HMAM and fully-associative caches are close to 83% and 50%, respectively. It shows that GParity behaves as a good protection code for detecting transient faults.

Table 2. Fault detection coverage with same hardware cost for 32Kbyte cache

parameters and results	4WSA	16HMAM	8HMAM	4HMAM	2HMAM	FA
# of protection bits	6	Variable	Variable	Variable	Variable	1
Radix of GParity	64	Variable	Variable	Variable	Variable	2
Avg. of calculated fault detection coverage	0.99	0.969	0.941	0.892	0.833	0.50
Avg. of miss ratio	0.197	0.1357	0.127	0.120	0.115	0.112

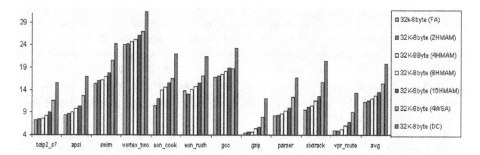

Fig. 3. Miss ratio (%) of fully-associative, several HMAM and direct mapped cache for various benchmarks

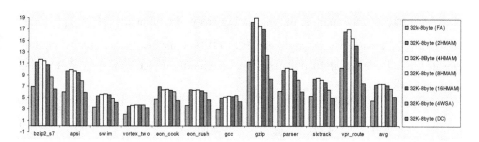

Fig. 4. Fault detection coverage per miss-ratio for direct mapped, 4-way set-associative, 2HMAM, 4HMAM, 8HMAM, 16HMAM and Fully-associative caches

Though fault detection coverages of the considered HMAM caches are less than set-associative cache, but their miss ratio is less than that of set-associative cache, as shown in the Fig. 3. Conversely, the fault coverage of the considered HMAM is more than fully-associative cache while their performance is less than it.

For more precisely consideration, we compared their fractions of coverage by miss-ratio as a good metric which incorporates both of fault-tolerance and perform-ance metrics. Designers like that fault coverage as well as hit-ratio to take high value. This leads designs to have a high value for the coverage per miss-ratio. Fig. 4 shows the coverage per miss-ratio for the mentioned cache architectures. As shown in the figure, the HMAM caches, specially in the average case, outperforms both of the fully-associative and 4-way set-associative cache.

In addition, as shown in Fig. 4, the 4HMAM cache has the best coverage per miss-ratio regarding to other considered HMAM caches. This implies that increasing the division factor in HMAM caches does not always increase the coverage per miss-ratio. Although it causes to increase the fault detection coverage, but in contrast, the miss-ratio would be increase, as well. By considering the fault detection coverage and miss-ratio trade-off, the 4HMAM cache is the best cache configuration for the given configuration.

5   Cost Analysis

In order to reduce latency of tag comparison in fully-associative caches, these memories are constructed using CAM (content addressable memories) structures. Since each CAM cell is designed as a combination of storage and comparison logic, the size of a CAM cell is about double as that of a RAM cell [14]. For fair *Performance/Cost* analysis, the performance and cost for various direct, set-associative and HMAM caches are evaluated. The metric used to normalize cost-area analysis is *rbe* (register-bit equivalent).

We use the same quantities used in [12], where the complexity of *PLA* (programmable logic array) circuit is assumed to be 130 *rbe*, a RAM cell as 0.6 *rbe*, and a *CAM* cell as 1.2 *rbe*. The *RAM* area can be calculated as [14]

$$\text{RAM} = 0.6 \Big[\text{\#entries} + \text{\#L}_{\text{sense amplifiers}} \Big] \cdot \Big[(\text{\#data bits} + \text{\#status bits}) + W_{\text{driver}} \Big] \tag{1}$$

where #entries is the number of rows in tag array or data array, $\text{\#L}_{\text{sense amplifiers}}$ is the length of a bit-line sense amplifier, #data bits indicates the number of tag bits (or data bits) of one set, #status bits is the state bit of one set, and W_{driver} is the data width of a driver.

The area of *CAM* can be given by [14]

$$\text{CAM} = 0.6 \Big[\sqrt{2} \cdot \text{\#entries} + \text{\#L}_{\text{sense amplifiers}} \Big] \cdot \Big[\sqrt{2} \cdot \text{\# tag bits} + W_{\text{driver}} \Big] \tag{2}$$

where # tag bits is the number of bits for one set in the tag array. The total area can be given by

$$\text{Area} = \text{RAM} + \text{CAM} + \text{PLA} \tag{3}$$

The area of HMAM cache was calculated by assuming that it is composed of several fully-associative caches, each of which has its specified size and tags. Table 3 shows *Performance/Cost* for various cache sizes. According to the Table 3, the HMAM cache (8HMAM, 4KB with 8-byte block size) shows about 40% area reduction compared to the conventional direct-mapped cache (DC, 8KB with 8-byte block size) while showing almost equal closed performance gains. Moreover, higher performance for HMAM scheme may be achieved by increasing the size of caches, compared to direct mapping schemes.

Table 3. Performance and cost of direct-mapped, 4-way set associative and HMAM caches

Cache Configuration	Area (*rbe*)	Avg. Miss ratio (%)	Avg. memory access time (cycles)
1K-8B (DC)	8168.1	46.87	5.6944
2K-8B (DC)	15491.1	40.34	6.4112
4K-8B (DC)	29840.35	33.82	7.4544
8K-8B (DC)	57934.45	29.34	8.4992
2K-8B (4WSA)	16125.9	35.64	6.7024
4K-8B (4WSA)	31089.5	30.21	5.8336
1K-8B (2HMAM)	8957.357	38.15	7.1040
8K-8B (2HMAM)	69635.27	22.02	4.5227
2K-8B (4HMAM)	17385.71	32.27	6.1632
4K-8B (8HMAM)	34218.59	28.39	5.5424

6 Conclusions

This paper proposed a fault detection scheme namely GParity and a cache architecture based on hierarchical Multiple associative mapping, called HMAM, which provides sets of different sizes. This architecture enabled designers to utilize the proposed protection code for every tag of cache line to improve fault detection coverage without compromising the performance or increasing significant area relative to the other cache schemes. Results obtained using a trace-driven simulator and soft-error injection revealed that HMAM can provide significant performance improvements with respect to traditional schemes and error detection coverage has been improved as compared with the already available single parity microprocessors.

References

1. Agarwal A., Hennessy J., Horowitz M.: Cache Performance of Operating Systems and Multiprogramming. ACM Trans. Computer Systems, Vol. 6, No. 4 , (1988) 393-431.
2. Agarwal A., Pudar S. D.: Column-Associative Caches: a Technique for Reducing the Miss Rate of Direct-Mapped Caches. Int'l Symp. on Computer Architecture (1993) 179-190.
3. Asadi G., Miremadi S. G., Zarandi H. R., Ejlali A. R.: Evaluation of Fault-Tolerant Designs Implemented on SRAM-based FPGAs. Proc. IEEE/IFIP Pacific Rim International Symposium on Dependable Computing, French (2004) 327-333.
4. Bertozzi D., Benini L., De Micheli G.: Low Power Error Resilient Encoding for On-chip Data Buses. Proc. of Design, Automation and Test in Europe Conference, France (2002) 102-109.
5. Brigham Young University: BYU Cache Simulator. http://tds.cs.byu.edu
6. Calder B., Grunwald D.: Predictive Sequential Associative Cache. Proc. 2nd Int'l Symp. High performance Computer Architecture (1996) 244-253.
7. Faridpour A., Hill M.: Performance Implications of Tolerating Cache Faults. IEEE Trans. on Computers, Vol. 42, No. 3 (1993) 257-267.

8. Farooqui A. A., Oklobdzija V. G., Sait S. M.: Area-Time Optimal Adder with Relative Placement Generator. Proc. of Int. Symp. on Circuits and Systems, Vol. 5, (2003) 141-144.
9. Imai H.: Essentials of Error-Control Coding Techniques. Academic Press, San Diego, (1990)
10. Karlsson J., Liden P., Dahlgern P., Johansson R., Gunneflo U.: Using Heavy-Ion Radiation to Validate Fault-Handling Mechanisms. IEEE Micro, Vol. 14 (1994) 8-23.
11. Kim S., Somani A.: Area Efficient Architectures for Information Integrity Checking in the Cache Memories. Proc. Intl. Symp. Computer Architecture (1999) 246-256.
12. Lee J. H., Lee J. S., Kim S. D.: A New Cache Architecture based on Temporal and Spatial Locality. Journal of Systems Architecture, Vol. 46 (2000) 1452-1467.
13. Miremadi G., Torin J.: Evaluating Processor-Behavior and Three Error-Detection Mechanisms Using Physical Fault Injection. IEEE Trans. Reliability, Vol. 44 (1995) 441-453.
14. Mulder J. M., Quach N. T., Flynn M. J.: An Area Model for On-Chip Memories and its Applications. IEEE journal of solid state Circuits, Vol. 26 (1991) 98-106.
15. Ranganathan P., Adve S., Jouppi N. P.: Reconfigurable Caches and their Application to Media Processing. Proc. Int. Symp. Computer Architecture (2000) 214-224.
16. Seznec A.: A Case for Two-Way Skewed-Associative Caches. Proc. Intl. Symp. Computer Architecture (1993) 169-178.
17. Shirvani P., McCuskey E. J.: PADded Cache: A New Fault-Tolerance Technique for Cache Memories. Proc. 17th IEEE VLSI Test Symp. (1999) 440-445.
18. Smith A. J.: Cache memories. Computing Survey, Vol. 14, No. 4 (1982) 473-530.
19. Standard Performance Evaluation Corporation: SPEC CPU 2000 benchmarks. http://www.specbench.org/osg/cpu2000
20. Intel Corporation: Pentium® Family Developer's Manual. http://www.intel.com
21. Reed R.: Heavy Ion and Proton Induced Single Event Multiple Upsets. IEEE Nuclear and Space Radiation Effects Conference (1997)
22. Swazey P.: SRAM Organization, Control, and Speed, and Their Effect on Cache Memory Design. Midcon/87 (1987) 434-437.
23. Zarandi H., Sarbazi-Azad H., "Hierarchical Binary Set Partitioning in Cache Memories," to appear in The Journal of Supercomputing, Kluwer Academic Publisher, 2004.
24. Zarandi H., Miremadi S. G., Sarbazi-Azad H., "Fault Detection Enhancement in Cache Memories Using a High Performance Placement Algorithm," IEEE International On-Line Testing Symposium (IOLTS), 2004, pp. 101-106.
25. Zhang W., Gurumurthi S., Kandemir M., Sivasubramaniam A.: ICR: In-Cache Replication for Enhancing Data Cache Reliability. In Proceedings of the International Conference on Dependable Systems and Networks (DSN) (2003) 291-300
26. Zhang C., Vahid F., Najjar W.: A Highly Configurable Cache Architecture for Embedded Systems. Int. Symp. on Computer Architecture (2003) 136-146.
27. Zhang C., Zhang X., Yan Y.: Two Fast and High-Associativity Cache Schemes. IEEE micro (1997) 40-49.

On the Effects of Errors During Boot*

Mário Zenha-Rela[1], João Carlos Cunha[2],
Carlos Bruno Silva[1], and Luís Ferreira da Silva[2]

[1] University of Coimbra, 3030-290 Coimbra, Portugal
{mzrela, cbsilva}@dei.uc.pt
[2] DEIS/ISEC, 3030-199 Coimbra, Portugal
{jcunha, lmferrao}@isec.pt

Abstract. We present the results of injecting errors during the boot phase of an embedded real-time system based on the ERC32 space processor. In this phase the hardware is initialized, and the processor executes the boot loader followed by kernel initialization. For this reason most system support is not yet available and traditional fault-injection techniques such as SWIFI cannot be used. Thus our study was based in the processor's IEEE 1149.1 (boundary-scan) infrastructure through which we injected about 5000 double bit-flip errors. The observations show that such system will either crash(25%) or execute correctly(75%), since only 2 errors eventually lead to the output of wrong results. However about 10% of faults originated latent errors dormant in memory. We also provide some suggestions on what can be done to increase robustness during this system state, in which most fault-tolerance techniques are not yet setup.

Keywords: dependability evaluation, embedded systems, fault-tolerance, fault-injection, boundary-scan.

1   Introduction

Reset is the most common error-recovery mechanism present in embedded computer systems. When some non-permanent error is detected, a simple hardware or software module triggers a reset that has the ability to bring the system from an erroneous state into an error-free state. This last-resort technique is used from the smallest embedded device (e.g. smart cards, mobile phones) to complex computer control systems provided with high degrees of redundancy to detect and tolerate a large class of errors [1].

After a reset the system is assumed to be clean from errors (both detected and latent) and may resume execution, rolling back to a previous state or jumping forward into a new one. However, bringing a system from a hard reset into a fully operational state implies a long series of complex and sensitive operations,

---

* This work was partially supported by the R&D Unit 326/94 (Center for Informatics and Systems, CISUC), and the Portuguese Agency for Innovation (AdI) through project BSCAN4FI.

C.A. Maziero et al. (Eds.): LADC 2005, LNCS 3747, pp. 131–142, 2005.

where the occurrence of a fault can lead to immediate drastic effects or stay latent until much later, with a high potential to induce a system failure.

This point is exacerbated by the fact that the dependability issues of most critical systems are based on a single failure model or that the mean time between failures (MTBF) is much longer than the recovery process. This requirement is mandated both by economic reasons and to make the dependability issues tractable.

However in many circumstances faults occur in 'bursts', i.e. they are clustered in the time domain: while the MTBF may be large the occurrence of successive faults (the phenomenological cause that originates errors [2]) may be very close followed by long periods of inactivity. This problem is particularly acute in space, where cosmological events such as solar flares may affect an on-board satellite computer during a recovery [3]. In most situations this problem can be acceptably managed as a single 'long' fault. In such cases the system restarts operating as soon as the disturbance intensity goes down a specified threshold. However, in many situations the fluctuations of the disturbance can lead to successive nearby non-overlapping faults affecting the computer system.

The problem of dealing with multiple errors can be handled very satisfactorily by resetting the system: after a hardware reset the system is considered to be in an error-free state so the potentially multiple errors are simply wiped out. However, if an error generated during boot manages to pass undetected, it may not be sufficient to simply reboot the system until all tests pass (boot is not an atomic activity). Then, if another error occurs in operation, the consequences may be dramatic as the system may not be able to handle a multiple error situation.

To the best of our knowledge this problem has never been addressed in the literature, most probably due to the lack of proper tools. During boot the system kernel is not loaded, device handlers and monitoring software are not yet operational. Thus, the use of dedicated hardware monitoring tools is mandatory, but the complexity of modern processors makes this approach unfeasible or extremely complex. A recently proposed fault-injection approach based on the IEEE 1149.1 (boundary-scan) standard [4] associated with on-chip debugging facilities seems to overcome this problem since it is orthogonal to the chip functionalities, and is permanently available whenever the chip is powered [5, 6].

This paper presents the undergoing research aiming at clarifying the effects of transient faults that occur during a system boot. Permanent faults are not considered in this study, since they can be more easily detected by diagnostics hardware.

The remainder of the paper is organized as follows: in the next section we preset the methodology used in this study, namely the testbed, the workload, the faultload and the measurements that were performed. In section 3 the activities performed during the boot sequence that eventually lead to the application launch are described. Section 4 contains the experimental results and in the following section possible ways to avoid failure are discussed. Section 6 closes the paper.

2   Experimental Methodology

Figure 1 depicts the main entities involved: the target system with the workload burned into its ROM, the fault-injection controller, the access to the target system boundary-scan port through its JTAG interface, and the database with the faultload and the outcomes.

The target hardware is the single board computer eVAB695 [7], built around the TSC695F (ERC32) 32-bit RISC space processor implementing the SPARC V7 architecture specification. This board includes 512K of radiation-hardened Flash ROM and 4096K of parity protected SRAM.

The TSC695F processor includes an integer unit (IU), a floating point unit (FPU), a memory controller (MEC) and a direct memory access (DMA) arbitrer. It also includes a watchog, two timers, an interrupt controller, one parallel and two serial interfaces. It supports on-chip debugging and boundary-scan testing accessible through the JTAG interface connected to a test access port (TAP) in the eVAB695. The TSC695F development was supported by the European Space Agency (ESA) aiming the space environment. This processor is currently on board the International Space Station and has also been adopted by the Brazilian and Chinese Space Agencies.

The application `gravity_v1.2`, is a program that calculates the trajectory of a mass (e.g. a satellite) attracted by a bigger mass (e.g. a planet) using Newton's gravity law. This workload, used extensively by ESA for testing purposes, outputs the successive satellite positions, in x-y coordinates.

The experimental workload `gravity` runs on top of the `RTEMS` `v4.6` kernel [8] ported to this specific hardware. RTEMS is an open source kernel designed to support applications with real-time requirements while being compatible with

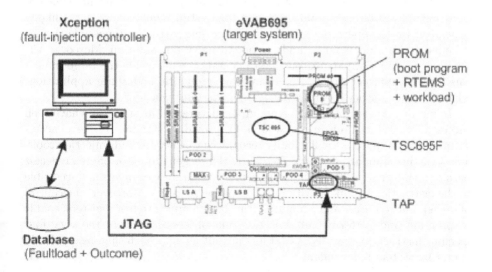

Fig. 1. The experimental testbed

open standards, namely the POSIX 1003.1b API and TCP/IP. The development of the board support package for the eVAB695 has also been supported by ESA and is available for download at the ESA web site [9].

The TSC695F was designed to stand the harsh space environment, so a number of error detection capabilities are built into the hardware, namely parity in the internal registers and data buses [10]. Thus, if we used the common single-bit fault model, this would generate easily detectable parity errors. Instead we adopted an adjacent double-bit flip error model to emulate SEU (*single event upset*) transients generated by cosmic radiation. In practice, this means that we are emulating bit-flips induced by the more energetic SEUs capable of flipping two adjacent bits.

The fault trigger is temporal: faults are injected following a uniform distribution during the boot phase, i.e. from the first clock cycle after reset to the moment the scheduler starts executing the first user application instruction. During this time frame we disturbed only processor registers (IU, FPU, control and status) since these could also emulate memory faults (e.g. erroneous values copied from ROM to RAM or into a wrong memory address). Moreover, the memory and buses are parity protected.

The injection of faults during system startup is not easily achieved with traditional techniques. Processor pin-level fault injection is currently an unfeasible option due to the ever-growing pin-hidden operations (e.g. prefetching and internal caches), as well as high clock frequencies. Radiation induced techniques, although applicable during system startup, pose known limitations of location and time control. Software induced fault injection (SWIFI) offers a level of control and efficiency hardly achievable by other techniques targeting processors. However, its dependence on specific routines that use the operating system resources while reacting to programmed breakpoints makes it unusable during system startup, the time frame focused in this study. In addition, practical SWIFI implementations have injection cores and operation (e.g. setup fault, collect data) highly dependent on the operating system design turning virtually impossible the development of an independent and pre-operating system solution. Moreover, the presence of potentially dangerous instrumentation code inside the target makes this kind of approaches less interesting for aeronautic and space applications where it is mandatory to 'test what you fly, fly what you test'.

In recent years the boundary-scan infrastructure and its successors have been successfully used for fault injection [5, 6, 11] providing standard low level access without giving up from the flexibility recognized to software fault injection tools. Through this standard test interface port, the target processor offers an access path to its internals, allowing injecting faults even in state elements not accessible to the instruction set, like parity bits or pipeline registers. Moreover, the control of breakpoint resources and running status, both mapped to test registers, enable to program and perform fault injection and observations from the very first machine instruction executed, i.e. at any instant, an approach that is completely operating system independent.

The fault injection campaigns presented in this study were performed using an improved version of Xception^{TM} [12, 13]. This is an automated fault injection environment designed to accommodate a variety of fault injection techniques namely the target processor on-chip debugging facilities available through the standard boundary-scan infrastructure.

The metrics collected were devised to provide a meaningful view of the target system robustness in face of faults injected during the boot phase:

Crash - the processor halted or was trapped in an endless loop. A hardware reset was needed to resume the experiments, so the watchdog timer could reboot the system.

OK/Clean - the system terminated correctly the boot sequence, the application was launched, and no errors were observed neither in the kernel nor in the application outputs. This involved a full scan of the target system memory segments (text, data, heap and stack) and processor registers after the boot and when the application terminated. We also collected the boot execution time in clock cycles.

OK/Latent - the boot sequence finished, the application was launched and no errors were observed in the outputs produced. However, internal (latent) errors which did not come to light during the experiment duration were present in the memory effectively used (errors in unused memory areas were not considered).

Wrong - the boot sequence finished and the application was launched but terminated with errors in the outputs produced.

The experiments were much simplified because the workload was being run in a controlled environment so the system state was deterministic. Through the boundary-scan port we could freeze the processor to modify and collect system data. Nevertheless, due to the low bandwidth of the IEEE 1149.1 interface each single injection run lasted more than 5 minutes, which meant that the experiments have taken several weeks running unattended.

3   The Boot Sequence

When a computer system is powered-on a long series of sensitive events occur before the target applications starts executing. These events aim at checking if hardware components are functional, configuring them, and loading software from a non-volatile storage media (ROM, flash RAM or hard disk) into main memory. Non-trivial embedded systems normally make use of a kernel which, after being loaded, must run through a complex initialization process. While this sequence of events is system specific, it usually follows these major steps (Fig. 2):

1. Power-on self test (POST) — when the processor is turned on the hardware performs a built-in self-test and some registers are initialized to a default value, namely the program counter (PC). Its default value points to a fixed

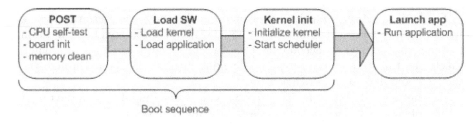

Fig. 2. Boot sequence

memory address in ROM where is located the very first machine instruction executed by the processor. The subsequent tasks are performed under software control: configuration and status registers are initialized, interrupts are disabled and common registers (IU and FPU) are cleared. Board specific code detects the hardware configuration (e.g. the number of serial ports and the top of memory to initialize the stack), performs diagnostics to check if the basic components are in perfect condition (e.g. test and clear RAM memory), and initializes some hardware components (e.g. I/O ports).

2. Kernel and application load — as happens in most embedded systems, the kernel is loaded as an application library and the different segments (text and initialized data) of the kernel and workload are copied from ROM into the RAM areas. Usually the ROM images are compressed so the copy also involves decompressing those segments. If this software resides on disk, a loader application is first copied from ROM and then loads the kernel and workload.

3. Initialize and start kernel — the kernel data structures are initialized, the most complex parts of the hardware are configured (e.g. co-processor, if present) and the device drivers are installed. Finally, the kernel scheduler starts executing by enabling interrupts.

4. Launch application — the 'main' routine in the user application code is entered and starts executing.

In figure 3 we present the *time × space* execution profile of the boot sequence for the target workload used in our experiments with *time* (clock cycles) in the horizontal axis and the *address* of instructions executed in the vertical axis.

Instead of being spread all over the address space, the memory accesses seem almost continuous (dark) horizontal bars. This indicates a tight access locality of the machine instructions executed, i.e. most of the system activity is centred in very few lines of code that are either clearing the memory or decompressing the application segments from ROM into RAM. We can see that the boot program starts executing in the ROM (lower addresses) followed by a long (about 4 million clock cycles) clearing of RAM memory. Then, the application and kernel code are copied into RAM (another 3 million clock cycles) followed by the initialized and uninitialized data areas (the 'uninitialized' data areas are effectively initialized to zero). This code fragment is loaded into RAM because fetching this code directly from ROM would be much slower. This temporary area is located near the

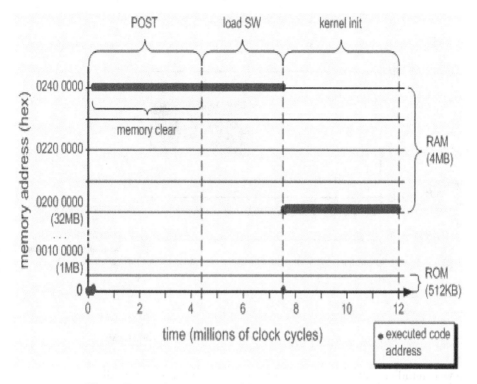

Fig. 3. Trace of instruction addresses accessed during boot

RAM top (Fig. 4) in the future stack area prior to the stack initialization, thus not conflicting with the boot operations being performed. Finally, the different kernel initialization routines are called to prepare the application launch by the scheduler (Fig. 3, 'kernel init').

The boot duration depends on the application size: the larger it is, the longer it takes to reboot. Moreover, if a large number of libraries were used, the longer it would be. In such embedded systems, services which are not required by the application are not even linked into the ROM image (e.g. if the workload did not use real arithmetic the floating point libraries would not be loaded).

In the presented testbed the kernel initialization code starts execution around the $7.500.000^{th}$ clock cycle. This means that for about 70% of the boot time the hardware is performing extremely tight code (cycles of about 10 machine instructions). As we shall see later this has a direct impact on the system's behaviour under faults.

Finally, around the $12.000.000^{th}$ clock cycle after reset, the application is started. At 20MHz clock frequency this means that 0.6 seconds are required for boot. While this seems a negligible fraction of time in missions lasting for decades, an error occurring during a reboot can have a dramatic impact on dependability, since it is manipulating extremely sensitive parts of the system, such as memory (code and data segments), pointers, kernel data structures,

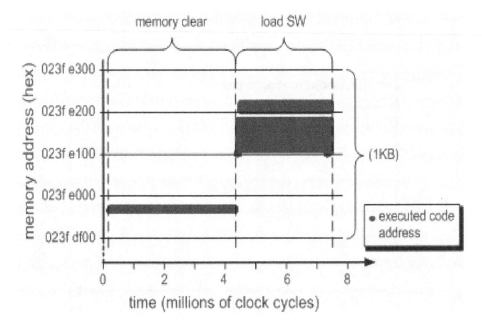

Fig. 4. The initial boot code is executed near the RAM top

device handlers, hardware configurations, etc. Furthermore, all this functions are executed in privileged mode, and for most of this time no error handlers are active and hardware based fault-tolerance support may not be configured yet.

4   Experimental Observations and Discussion

Table 1 presents an overview of the target system behaviour after the injection campaign involving 4997 effectively injected faults. It depicts the final system behaviour (columns) versus the system state observed when boot terminated (rows).

The most significant observation is its resilience to failure: the system either produces correct outputs (OK/75%) or no output is generated at all due to crash(25%). A residual 2 faults lead to the production of wrong outputs. It must be stressed that this behaviour is clearly distinct from what we have observed in previous research dealing with faults injected during steady-state operation in similar embedded systems (figures were around Clean(50%), Crash(48%), Wrong(2%) [14, 15]).

As would be expected the observations show that every OK/Clean outcome arises from a clean boot environment. The large number of samples where the system was unaffected (65%) indicates the presence of a significant intrinsic hardware redundancy. In fact, as was referred in section 3, during about 70% of the boot time the processor is performing extremely tight code (cycles of about 10 machine instructions, using only 5 of its 256 windowed registers), checking

Table 1. Overview of the target system behaviour

State after boot	$\sum$	Crash	OK/Clean	OK/Latent	Wrong
Clean	3250	0	3250	0	0
Corrupted	490	1	0	487	2
Crash	1257	1257	0	0	0
Totals	4997	1258 (25%)	3250 (65%)	487 (10%)	2 (0.04%)

and clearing memory (POST) and moving the data from ROM into the RAM space. This means that most of the processor resources are effectively idle and thus unused, which explains its resiliency to failure. Whenever a resourceful state machine (such as a processor) uses very few of its resources, the probability of a disturbance affecting the active functional units is reduced. This observation agrees with previous observations on the correlation of system load and the occurrence of errors [16].

Crashes are dominant from errors occurring during the transfer of the application image from ROM into RAM (Fig. 5, 'load SW'). Note that the decompression of the RAM image is parity protected, but since we are injecting adjacent double bit flip-faults, this mechanism is not enough to prevent the corruption of the memory image. It was observed that only one fault eventually leading to crash managed to reach the application entry point. This fault corrupted a global register (used as frame pointer) during the kernel initialization. The remaining 1257 faults crashed the system when the corrupted kernel code was executed, so the boot phase never terminated.

About 10% of all faults (490) lead to a corrupted system after boot termination and to the presence of (487) latent errors despite the production of correct outputs. The characterization of these errors show a prevalence of faults injected during the final phases of the boot, i.e. during the kernel initialization. These errors were resident in kernel structures which have not been used.

A most undesirable behaviour of any computer system is the production of wrong outputs without being detected by any error detection mechanism being rather preferable a crash (no outputs produced). This is known as the fail silent model [17], an assumption under which most dependable systems are designed. The fail-silent behaviour is usually associated with the evaluation of dedicated error detection mechanisms [14].

By tracing the executions that generated wrong outputs we observed that these two faults corrupted fixed data areas (static data) during the memory initialization phases. This behaviour agrees with the research performed on [15, 18] on the resiliency of errors and checkpoint corruption.

The analysis of the error impact versus fault profile showed a slightly higher correlation between the trigger address, i.e. what the processor was doing at the injection instant, rather than a specific target register. Obviously corrupting the PC lead to crash and the most sensitive address was at the long segment transfer routines, but beyond these exceptional cases no particular dependency was observed.

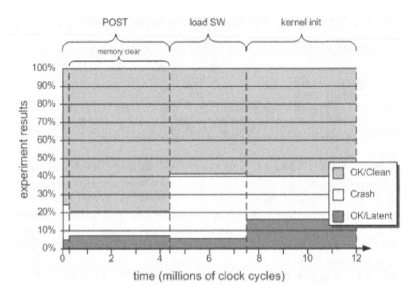

Fig. 5. Distribution of fault outcomes along the boot phases

5   Tolerating Boot Errors

Based on the previous observations we shall now suggest possible ways to tolerate errors that may arise during boot.

5.1   Detection of Timing Deviations

The initial boot phase is a deterministic sequence that is not interrupted or subject to different execution threads. In fact, since interrupts are disabled, there are no external events that could cause diverse control flow sequences. The processor itself does not have any indeterministic characteristic, such as speculative execution. Interrupts and thus the scheduler are only enabled at the very end of boot. This means that we can know precisely the boot sequence duration, be it in real time or clock cycles. Since a clean environment after boot implies a correct timing, then the boot duration may be used for error detection. Effectively the observations show that about 13% of latent errors are associated with incorrect boot timing.

These observations have immediate applicability for error detection purposes: if a watchdog timer is set to the boot duration (plus a minor margin, since it is asynchronous relative to the processor clock), and a small routine at the end of boot checks the watchdog counter for an early boot termination, they will detect every crash and 13% of boot corruptions leading to latent errors (despite the fact that these would seem like false alarms since the outputs would be correct). In the current target, up to 62 out of 487 latent errors would have been detected. The two samples leading to the production of wrong outputs would not be detected since they showed a correct timing.

5.2 Detection of State Corruption

Since the boot sequence is fully deterministic, we can collect the system state in advance. Later in operation, during boot, the full system state (memory and relevant hardware configuration registers) is checked for corruption using (e.g.) a resident CRC check, and if any deviation is found a reboot is forced. The point is that in systems where there is no memory protection (as is the case) the check itself can corrupt memory areas which have already fed the CRC calculation. This suggests that –despite the additional power required and the time execution overhead– the use of ROM for the code segments and fixed data areas should be considered. Alternatively there could be some form of blocking the write accesses to such 'fixed content' areas.

It is mandatory to complement this approach with the watchdog timer referred above, since the CRC check itself may be bypassed by a control flow error. If, for some reason, this check were not executed, then the watchdog timer would expire and force a reset.

6 Conclusion

In this paper we presented an experimental study on the behaviour of an embedded real-time system under the occurrence of faults during boot. During this time frame operating system support is not yet available and traditional fault-injection techniques cannot be used. Thus, our research was based in a fault-injection approach based in the IEEE 1149.1 (boundary-scan) standard, since this infrastructure is orthogonal to the chip functionalities, and is permanently available whenever the chip is powered.

The activities performed during boot show that the system either produces correct outputs (75%) or no output is generated at all due to crash (25%). However, about 10% of the faults caused latent errors in the system despite the production of correct outputs. Only 2 faults lead to the production of wrong outputs.

The insights achieved from this study provide clues on what can be done to increase robustness during this system state in which most fault-tolerance techniques are not yet setup. The determinism of the boot, both in time and in the actions performed, indicate that the boot resiliency to failure can be significantly increased by i) a watchdog timer finely tuned to the boot duration, ii) preventing writes to addresses with fixed contents and iii) associating a watchdog timer to a CRC check of the system memory and relevant hardware configuration registers.

Acknowledgements

We acknowledge Professor Algirdas Avižienis as the source of inspiration for this work as he remarked during the EDCC4 conference that no SWIFI tool could be used to inject faults immediately after a reset.

References

[1] J. Cunha, A. Correia, J. Henriques, M.Z.-Rela, J. Silva, *Reset-Driven Fault Tolerance*, 4th European Dependable Computing Conference (EDCC-4), Toulouse-France, October 23-25 2002, LNCS 2485, A. Bondavalli, P. Thevenod-Fosse (Eds.), Springer-Verlag Heidelberg 2002, pp. 102 - 120.

[2] J.-C. Laprie, A. Avižienis, H. Kopetz (Eds.), *Dependability: Basic Concepts and Terminology*, Springer-Verlag, ISBN:0-3878229-6-8, 268 pages, New York 1992.

[3] S. Potteck, *La conception de systèmes spatiaux*, Éditions du Schèmectif, Juillet 2001, ISBN 2-9513724-0-X (2 Tomes).

[4] IEEE Std 1149.1-2001, *IEEE Standard Test Access Port and Boundary-Scan Architecture*, ISBN: 0738129445, New York, 2001.

[5] P. Folkesson, S. Svensson, J. Karlsson, *A comparison of simulation based and scan chain implemented fault injection*, In Proc. of 28th Symposium on Fault Tolerant Computer Systems (FTCS-28), Munich, Germany, IEEE Computer Society 1998, pp. 284-293.

[6] L. Santos, M.Z.-Rela, *Constraints on the use of boundary-scan for fault injection*, in Proc. First Latin-American Dependable Computing Symposium, S. Paulo, Brazil, Oct. 2003, Lecture Notes in Computer Science, LNCS 2847, Springer-Verlag Heidelberg 2003.

[7] TSC695 Evaluation Board User Guide Manual, Rev.C 01/00, ATMEL Corp.2000 http://www.estec.esa.nl /microelectronics /presentation/ERC32.pdf

[8] RTEMS: *Real-Time Executive for Multiprocessor Systems* http://www.rtems. com/

[9] http://www.estec.esa.nl/wsmwww/erc32/freesoft.html

[10] J. Gaisler, *Evaluation of a 32-bit Microprocessor with Built-In Concurrent Error-Detection*, in Proc. FTCS-27, June 25-27, IEEE Computer Society 1997, pp. 42-46.

[11] P. Yuste, J.-C. Ruiz, L. Lemus, P. Gil, *Non-intrusive Software-Implemented Fault Injection in Embedded Systems*, in Proc. First Latin-American Dependable Computing Symposium, S. Paulo, Brazil, Oct. 2003, LNCS 2847, Springer-Verlag 2003, pp. 23 - 38.

[12] Xception^{TM}-*Enhanced Automated Fault-Injection Environment*, 2002, http://www.xception.org.

[13] J. Carreira, H. Madeira, J.G. Silva, *Xception: A Technique for the Experimental Evaluation of Dependability in Modern Computers*, IEEE Trans. on Software Engineering, February 1998.

[14] H. Madeira, J.G.Silva, *Experimental Evaluation of the Fail-silent behaviour in Computers without Error Masking*, In Proc. FTCS-24, Austin-USA, IEEE Computer Society 1994, pp. 350-359.

[15] J. Cunha, R. Maia, M. Z.-Rela, J.G. Silva, *A Study of Failure Models in Feedback Control Systems*, in Proc. DSN'2001, July 1-4, 2001, Göteborg-Sweden, IEEE Computer Society 2001.

[16] R. K. Iyer, D. Tang, *Experimental Analysis of Computer System Dependability*, Chap. 5 in Fault-Tolerant Computer System Design (ed. D.K. Pradhan), ISBN 0-13-057887-8, Prentice Hall 1996, pp. 282-392.

[17] D. Powell, G. Bonn, D. Seaton, P. Verissimo, *et. al*, *The Delta-4 approach to dependability in open distributed computing systems*, in Proc. FTCS18, Japan, June 1988.

[18] J. Vinter, A. Johansson, P. Folkesson, J. Karlsson, *On the Design of Robust Integrators for Fail-Bounded Control Systems*, DSN2003, ISBN 0-7695-1952-0, IEEE Computer Society 2003, pp. 415-424.

A Fault Tolerant Approach to Object Oriented Design and Synthesis of Embedded Systems

M. Fazeli, R. Farivar, S. Hessabi, and S.G. Miremadi

Department of Computer Engineering,
Sharif University of Technology,
Azadi Street, Tehran, Iran
{m_fazeli, r_farivar}@ce.sharif.edu
{hessabi, miremadi}@sharif.edu

Abstract. The ODYSSEY design methodology has been recently introduced as a viable solution to the increasing design complexity problem in the ASICs. It is an object-oriented design methodology which models a system in terms of its constituting objects and their corresponding method calls. Some of these methods are implemented in hardware; others are simply executed by a general purpose processor. One fundamental element of this methodology is a network on chip that implements method invocation for hardware-based method calls. However this network is prone to faults, thus errors on it may result into system failure.

In this paper an architectural fault-tolerance enhancement to the ODYSSEY design methodology is proposed which covers this problem. It detects and corrects all single event upset errors on the network, and detects all permanent ones. The proposed enhancement is modeled analytically and then simulated. The simulation results, while validating the analytical model, show very low network performance overhead.[1]

1 Introduction

The potential ASIC design complexity has grown at a rate of 58% per year for the last two decades, while the designer productivity at the same time has only raised 21% per year [1]. This has led to a growing design productivity gap between manufacturing capability of chips and the functionality that designers can implement in unit time. The manufacturing capability is predicted to grow at the same rate for another decade and hence the gap must be filled by increasing designer productivity rate.

To fill in the gap, new design methodologies and tools are quite necessary. Object-oriented design, successfully used for several years by the software community, is a rather different approach to complexity management compared to traditional hardware design methodologies that has received much attention recently. The Object Oriented methodology suggests modeling the system in terms of its constituting data objects and their corresponding method calls, while traditional methodologies concentrate on structurally decomposing the target architecture of the system.

---

[1] This work is supported by a research grant from the Department of High-Tech. Industries, Ministry of Industries and Mines of the Islamic Republic of Iran.

C.A. Maziero et al. (Eds.): LADC 2005, LNCS 3747, pp. 143–153, 2005.

As an example of the usage of object oriented design methodology in hardware design, the ***Object-oriented Design and sYntheSiS of Embedded sYstems (ODYSSEY)*** methodology has been recently introduced [2, 6]. It is a system-level synthesis methodology for embedded systems that begins from an object-oriented code in C++ and synthesizes it into an Application Specific Instruction-set Processor (ASIP) and software that runs on it. Virtual-method calls are implemented as packets sent over a ***Network on Chip (NoC)*** [5] from the caller module to the called one, carrying the parameters of the call as the packet data payload. The Functional Units (FU) addresses and object numbers are assigned such that routing of a packet is equivalent to dynamic binding of the corresponding virtual-method call.

As technology scales, however, errors and faults are becoming increasingly common in the NoCs. Crosstalk interferes with signal transmission, while soft errors result in random bit-flips throughout the design [3]. Critical leakage currents and high field effects will also lead to more transient and permanent failures of signals, logic values, devices, and interconnects [4]. Since the area resources available on the chip are limited, implementing traditional fault tolerant algorithms and architecture such as modular redundancy in the NoC domain is infeasible. Therefore, other techniques must be developed if fault tolerant NoCs are to become possible. Previous work in this area is limited [7, 8, 9] as NoC design is still in its infancy. Guerrier et al. [10] presented a NoC design called SPIN that was based on fat-tree topology. They also presented the router architecture and cycle accurate performance model for their NoC design. Sgroi et al. [11] discussed a platform based SoC design methodology that proposed the inclusion of NoC for supporting on-chip communication. Dally et al. [12] demonstrated the feasibility of the NoC and estimated that the NoC places an area overhead of 6.6%. Benini et al. [13] in their conceptual paper on NoC, predict that packet switched on-chip interconnection networks will be essential to address the complexity of future SoC designs. Kumar et al. [14] presented a conceptual system-level architecture that allowed a mesh-based NoC to accommodate large resources such as memory banks, FPGA areas, or high performance multi-processors. Note that in the nanoscale regime, crosstalk in long global communication wires is expected to be the major source of errors. [2]

It should also be noted that this problem gets worse in ODYSSEY design methodology since there is a great probability of system failure in case of an error in the packets traversing the NoC between functional units and the traditional CPU in the ODYSSEY methodology, since these packets contain the functional units addresses and object numbers.

In this paper, a fault tolerant architectural enhancement to the ODDYSEY methodology is proposed, which adds transient NoC fault detection and tolerance, as well as permanent fault detection, and is based on the principle of packet re-transmission. The idea is to introduce redundancy into the system by sending each packet twice, the original packet is sent first and the complement of the packet is sent next. However, the amount of time overhead is far lower than what one might predict first, because of the potential overlapping of the second transmission time with the processing time in the called FU. The simulation results show very negligible overhead figures in the abundance of faults, while not exceeding 51% overhead with a 100% fault probability in each packet transmission. In addition, the amount of imposed hardware is rather low, as just a buffer stage and a comparator and two very simple state machines are required per FU. On the other hand, the transient fault detection coverage and tolerance for the Single Event Upset (SEU) transient fault model is 100%. All of the permanent faults in the NoC will also be detected.

The structure of this paper is along these lines: following the introduction the ODYSSEY design methodology will be introduced in section 2. Section 3 discusses the proposed fault tolerance architectural enhancement and an analytical model for the time overhead will be obtained in Section 4. The simulation results for time overhead of the proposed scheme are shown in Section 5. Finally, Section 6 concludes the paper.

2 ODYSSEY Design Methodology

The ODYSSEY design methodology is a system-level synthesis methodology for embedded systems that starts from an object-oriented (OO) system model and implements it as an application specific processor, called Object-Oriented Application-Specific Instruction Processor (OO-ASIP), and the software for it. A method of a class is the minimum unit that is assigned to either hardware or software partition and is accordingly called *hardware* or *software method*. To efficiently dispatch virtual method calls to hardware as well as software methods, a network-based mechanism has been proposed [5] that dispatches virtual method calls as packets sent over an on-chip network to which all possible functional units are connected. The method dispatching via network routing is identified by the inherent NoC architectures; this realizes polymorphism for free and is used in this research.

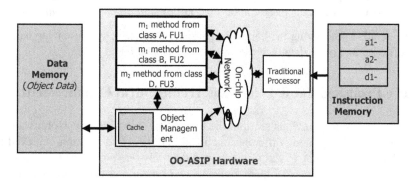

Fig. 1. The internal architecture of the OO-ASIP. Note the NoC connecting the functional units and the traditional processor.

The embedded system architecture is depicted in Figure 1. The system is a Network-on-Chip (NoC) architecture that consists of a processor core along with a set of hardware functional units. The architecture is specifically designed to suit object-oriented (OO) applications.

A typical OO application defines a library of classes, instantiates objects of those classes, and invokes methods of those objects. The implementation approach for each of these three major components of an OO application is described below. For presentational purposes, we follow the C++ syntax in describing each component.

- **Class library:** Each class consists of variable declarations and method definitions. Variable declarations are compile-time information and do not require a corresponding component in the implementation. Methods of the OO class library are either implemented in software or in hardware (e.g. A->m_1, B->m_1 and D->m_2 are implemented in hardware).
- **Object instantiations:** these are specified in the *main()* function. A memory portion should be reserved for each instantiated object to store the values of its data items. This memory portion is allocated in a data memory (the gray box at the left-hand side of Figure 1) that is accessible to the processor core as well as all FUs.
- **Method invocations:** the sequence of method invocations is specified in the *main()* function of the application. The executable code of this function comprises another part of the instruction memory (see Figure 1). The processor core starts by reading the instructions specified in the *main()* function of the application. Whenever a method call instruction is read, the corresponding implementation is resolved and invoked. This may result in calling a software routine or activating an FU (e.g. A->m_1 in Figure 1). Each method implementation (be it in hardware or software) can also call other methods. Each method call is assumed as a network packet. Each method call is identified by a method, an object, and the parameters of the call; hence, the bit-field concatenation of these items represents the method call and comprises the packet to be sent; i.e. *<method-id.object-id.params>*.

The details on resolving method calls, passing parameters, synchronizing hardware and software, and other details of the architecture can be found in [2].

3 The Proposed Fault Tolerance Architectural Enhancement

As mentioned in the previous section, the ODYSSEY methodology exploits an NoC to distribute packets among different functional units. Polymorphism is achieved as a side effect of this design. It was also assumed that only one functional unit can access the network at each time; i.e, there could be one sender and one receiver functional units present on the network, the other functional units may be busy processing previous method calls, or may simply be free.

The fault model considered in this research is Single Event Upset (SEU). The occurrence of each fault on the NoC in this methodology will most probably result in a system failure due to the fact that a packet sent over the network has two fractions, one is the address of the caller and the called functional units and the other one contains some parameters; hence incorrectness of each data could cause result errors. To protect the system against such failures, a fault tolerant architectural enhancement is proposed which is based on the principle of packet retransmission. After the transmission of a packet, the complement of that packet is sent to the same receiver. Sending the complement of a packet is selected because it would rarely be possible that a transient SEU fault could alter the same bit in a packet and its complement in an unrecognizable manner. Also a permanent fault such as a stuck-at fault can also be detected. Since a packet contains the destination address field, the routing operation would get disturbed by complementing the packet in the second transmission. To address this issue an *inversion bit* is appended to each packet. For the original packet transmission this bit is cleared, so the routing logic of the network does not alter the address field, on the other hand during the transmission of the complemented packet the inversion bit; is set automatically, when the packet is complemented. The routing logic complements the address field of a transmitted packet when it encounters a set inversion bit so that the complemented packet can reach its correct destination. When the original packet and its complement are present in the destination functional unit, a comparison is performed. To avoid unbounded waiting for arrival of a complement packet an internal watchdog timer is included in the receiving portions of all functional units.

In the case of a mismatch, a dis-acknowledgement packet (*Nack*) would be sent and these steps would be repeated once more; otherwise the acknowledgement packet (*Ack*) would be sent to the sender functional unit. If there was a mismatch again in the second round, the occurrence of a permanent fault would be reported. It should be noted that the processing of the packet in the receiver functional unit starts by the end of transmission of the original packet; the transmission time of the complemented packet could be overlapped with the processing time of the original packet. In other words, if the processing time exceeds the transmission time of the complemented packet, which is obviously the case in OO-ASIP, there would be no time overhead in the absence of the faults (the normal case).

The state transition diagrams of the Moore FSMs of the control circuitry in the sending and receiving part of a functional unit are shown in Figures 2.a and 2.b. The *Ack*, *Nack* and *Packet_ready* are the input signals in the sending part FSM. The activation of the *T(p)* output signal triggers the transmission of an original packet and The activation of the *T(-p)* output signal triggers the transmission of a complemented packet. In the receiving part of a functional unit, *R(a)* and *R(-a)* input signals represent the reception of an original or a complement packet respectively. The *C* input signal shows the result of the comparison and *TO* input signal shows the time out event of the internal watchdog timer. The *TT* output signal resets and starts the internal watchdog timer, the *Ack* and *Nack* output signals are also generated in this FSM.

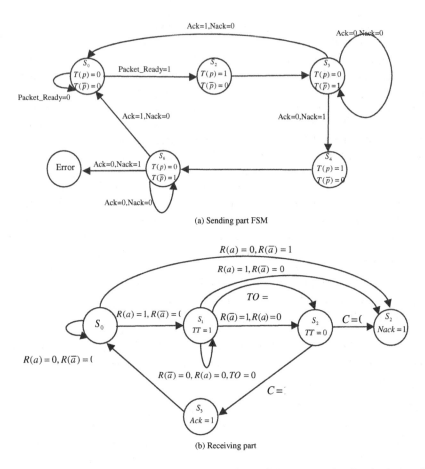

Fig. 2. The state transition diagram of the Moore FSMs of the control circuitry in the sending and receiving part of a functional unit

Timing diagrams of the sender and receiver functional units in the case of two different functional units, the first one with a one clock processing delay and the second one with a two clock processing delay, are depicted in Figure 3.a and 3.b. Note the concurrency of the complemented packet with the processing of the original packet.

Due to the nature of the SEU errors it is impossible for an SEU to corrupt both the original and the complement packet in an undetectable manner. By giving another chance of retransmission, all of the SEU errors can be tolerated. It is in the nature of ODYSSEY methodology that more complex methods of a class are implemented in hardware. The hardware overhead imposed by the presented scheme, which consists of a buffer, a comparator and a simple finite state machine (FSM) per functional units, is negligible compared to the complexity of the functional units.

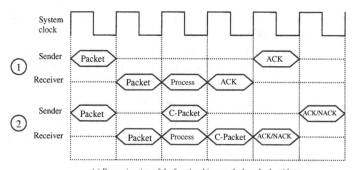

(a) Processing time of the functional is one clock cycle. 1- without
fault tolerant mechanisms 2- with fault tolerant mechanisms.

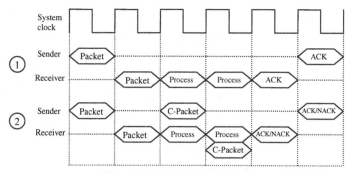

(b) Processing time of the functional is 2 clock cycles or more. 1- without
fault tolerant mechanisms 2- with fault tolerant mechanisms.

Fig. 3. Timing diagrams of the sender and receiver functional units

4   An Analytical Evaluation of the Time Overhead

As mentioned before the hardware overhead of the proposed scheme is negligible
compared to the complexity of the functional units. Furthermore the number of
transistor is no more a bottleneck in modern VLSI fabrication [1]. On the other
hand, the time overhead of a fault tolerance enhancement scheme is its most
important aspect. To extract the time overhead of the proposed scheme an
analytical model is introduced in this section. The following parameters are used in
the model:

- n : The number of available functional units.
- P_{fault} : The probability of fault incidence in the NoC.
- $P_{OneClock, \, i}$: The probability that functional unit i has a one-clock
  processing-time.

- α : The time overhead, in clock cycles, of a one-clock processing-time when a fault happens, this is equal to 7 in the proposed scheme.
- β : The time overhead, in clock cycles, of a two-clock or more processing-time when a fault happens, this is equal to 6 in the proposed scheme.
- γ : The time overhead, in clock cycles, of a one-clock processing-time in the absence of faults, this is equal to 1 in the proposed scheme.
- θ : The time overhead, in clock cycles, of a two-clock or more processing-time in the absence of faults, this is equal to zero in the proposed scheme.
- μ : Network overhead per packet transmission in the unprotected architecture, which is assumed to be 4 in the proposed scheme (according to figure 3).
- ξ_i : (Usage Factor) the probability of functional unit i being used in each method invocation.
- x_i : A stochastic variable that represents the processing-time, in clock cycle, of functional unit i.
- $\Gamma_i(x_i)$: The probability density functions of the x_i stochastic variable.
- $E(x_i)$: The expected value of the x_i stochastic variable, which can be computed as $E(x_i) = \sum_k x_i.\Gamma_i(x_i)$, where k is the range of values x_i can take.
- T : The imposed Time overhead

Using the preceding parameters the time overhead can be modeled by the following expression:

$$T = \frac{\sum_{i=1}^{n} [P_{fault}(P_{OneClock,i}.\alpha + (1-P_{OneClock,i}).\beta) + (1-P_{fault})(P_{OneClock,i}.\gamma + (1-P_{OneClock,i}).\theta)].\xi_i}{\sum_{i=1}^{n} [E(x_i) + \mu].\xi_i}$$

The first term of the addition operator in the numerator shows the extra clock cycles the proposed scheme necessitates when a fault happens, and the second term represents the overhead cycles in the absence of faults. The addition operator in the denominator demonstrates all of the clock cycles required in the transmission of an unprotected packet.

For the sake of simplicity, it is assumed that the usage factors ξ_i, the $P_{OneClock,i}$ and the $\Gamma_i(x_i)$ are the same for all functional units ($\xi_i = 1/n$, $P_{OneClock,i} = P_{OneClock}$ and $\Gamma_i(x_i) = \Gamma(x_i)$). Considering these simplifications and substituting the α, β, γ and θ parameters with their respective numerical values the following simplified expression can be deducted.

$$T = \frac{[P_{fault}(P_{OneClock}7+(1-P_{OneClock}).6)+(1-P_{fault})(P_{OneClock})]}{E(x)+4} = \frac{P_{OneClock}+6.P_{fault}}{E(x)+4}$$

It is evident from the latter expression that the time overhead is linearly proportional to the fault incidence probability and has inversely proportional to the distribution of functional units' processing-times. In the next section the correctness of this model is validated by simulation results.

5 Simulation Results

To validate the analytical model introduced in the previous section, a simulation of the NoC portion of an OO-ASIP based system was performed. The simulation results show that under transient fault injection in the NoC, the time overhead of the proposed architectural fault tolerance enhancement relies heavily on the distribution of the functional units' processing time. The results of the simulation for different fault probability values are depicted in Figure 4. To discover the time overhead three different scenarios were simulated with thirty functional units:

- In the best case, all of the functional units posses the maximum processing time allowable. As expected, in the best case there is no time overhead in the absence of faults, since retransmission of all the complemented packet is fully overlapped with functional unit processing times. In presence of faults the time overhead remains low because for transmitting each corrupted packet transmission an overhead of six-clock cycle would be imposed, and in the case of maximum processing time remains less than 20%. In this case $P_{OneClock}$ =0 and $E(x)$ =30 (maximum allowable clock cycle per FU in this simulation).

- All of the functional units had a processing time of one clock cycle in the worst case. This would be rare case in actual implementations. In this case the overhead does not exceed 140%. Note that the worst case that is simulated here is much exaggerated. A fault probability of one is not usually experienced in physical systems. In addition the functional units in an actual OO-ASIP implementation will consume more than one clock cycle to complete their operation. In this case $P_{OneClock}$ =1 and $E(x)$ =1.

- In the average case the functional units processing times had a uniform distribution between one clock cycle and the maximum processing time

- allowable. The highest time overhead in this case is 30% under a fault incidence probability of 1, which of course does not happen in physical systems. Thus the overhead figures in a physical system are much lower. In this case $P_{OneClock} = \frac{1}{30}$ and $E(x)=15.5$.

Due to the accurate matching of the extracted simulation and analytical results, it is apparent that the proposed expression models the system behavior precisely.

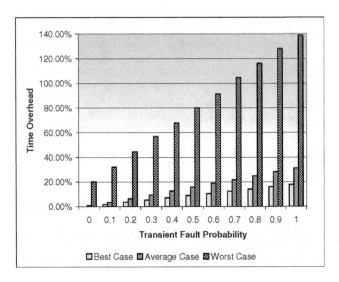

Fig. 4. The Simulation results demonstrating the Time Overhead vs. the transient fault probability

6 Conclusions and Future Works

An architectural fault-tolerance enhancement to the OO-ASIP design methodology was proposed in this paper. It is designed in such a way that it could detect and correct all Single Event Upset errors on the network, and could detect all permanent ones. The proposed enhancement was modeled analytically, the consistency of simulation and analytical results validate the model. The results also show very low levels of network performance overhead, hence suitability to be used in actual implementations.

The authors are currently working towards physical implementation of a few application specific systems with the enhanced, fault tolerant ODYSSEY methodology. When implemented, they will be objected to physical fault injection campaigns to evaluate the reliability of the proposed scheme.

Acknowledgement

Authors would like to acknowledge Maziar Gudarzi for his invaluable advices in the course of this research.

References

[1] International Technology Roadmap for Semiconductors (ITRS)-Design, 2001. http:// public. itrs. net/ Files/2002Update/2001ITRS/Design.pdf
[2] P. Vellanki et al. "Quality-of-service and error control techniques for mesh-based network-on-chip architectures", INTEGRATION, the VLSI journal 38 (2005) 353–382 380

[3] W. Dally and J. Poulton. *Digital Systems Engineering*. Cambridge University Press, 1998.

[4] Y. Taur and et. al. CMOS scaling into the nanometer regime. In *Proc. of the IEEE*, volume 85, April 1997.

[5] Goudarzi, M., Hessabi, S., Mycroft, A., "Overhead-free Polymorphism in Network-on-Chip Implementation of Object-Oriented Models," Proc. of Design Automation and Test in Europe (DATE'04), Feb. 2004, Paris.

[6] Goudarzi, M., Hessabi, S., "Object-Oriented Embedded System Development Based on Synthesis and Reuse of OO-ASIPs," Journal of Universal Computer Science (JUCS), In Press, Sep. 2004.

[7] T. Dumitras, S. Kerner, and R. Marculescu. Towards onchip fault-tolerant communication. In *Proc. Asia and South Pacific Design Automation Conference*, 2003.

[8] S. M. Hedetniemi, T. Hedetniemi, and A. L. Liestman. A survey of gossiping and broadcasting in communication networks. *NETWORKS*, 18:319–349, 1988.

[9] D.W. Krumme, G. Cybenko, and K. N. Venkataraman. Gossiping in minimal time. *SIAM J. Comput.*, 21(1):111–139, 1992.

[10] P. Guerrier, A. Greiner, A generic architecture for on-chip packet-switched interconnections, in: DATE, Paris, France, March 2000.

[11] M. Sgroi, M. Sheets, A. Mihal, K. Keutzer, S. Malik, J. Rabeay, A. Sangiovanni-Vincentelli, Addressing the system-on-a-chip interconnect woes through communication-based design, in: Proceedings of Design Automation Conference, June 2001, pp. 667–672.

[12] William J. Dally, Brian Towles, Route packet, not wires: on-chip interconnection networks, in: Proceedings of DAC, June 2002.

[13] Luca Benini, Giovanni De Micheli, Networks on chips: a new SoC paradigm, IEEE Comput. (2002) 70–78.

[14] S. Kumar, A. Jantsch, M. Millberg, J. Oberg, J.P. Soininen, M. Forsell, K.T.A. Hemani, A network on chip architecture and design methodology, in: IEEE Computer Society Annual Symposium, on VLSI, Pittsburg, Pennsylvania, April 2002.

Scheduling Fixed-Priority Hard Real-Time Tasks in the Presence of Faults

George Lima[1] and Alan Burns[2]

[1] Distributed Systems Lab (LaSiD),
Department of Computer Science (DCC),
Federal University of Bahia (UFBA),
Salvador, BA, Brazil
gmlima@ufba.br
[2] Real-Time Systems Research Group,
Department of Computer Science,
University of York, York, UK
burns@cs.york.ac.uk

Abstract. We describe an approach to scheduling hard real-time tasks taking into account fault scenarios. All tasks are scheduled at run-time according to their fixed priorities, which are determined off-line. Upon error-detection, special tasks are released to perform error-recovery actions. We allow error-recovery actions to be executed at higher priority levels so that the fault resilience of the task set can be increased. To do so, we extend the well known response time analysis technique and describe a non-standard priority assignment policy. Results from simulation indicate that the fault resilience of the task sets can be significantly increased by using the proposed approach.

1 Introduction

Hard real-time systems are those that *have* to produce correct results within specified deadlines. A flight control system is an example of such a system. Should it fail to produce correct or timely results, an accident may happen. In other words, high costs, in terms of human lives or monetary loss, are usually associated with failures in such a kind of system.

Due to the criticality level of their computation, dealing with hard real-time systems is not simple. In order to provide fault tolerance, the system must be designed making use of redundant components. In order to provide timeliness, the system computation must be organized so that its timing specifications are met. Also, there must be ways of proving the system timeliness given the characteristics of both the system and the environment it is subject to, which must take the presence of faults into account. Preferably, this timeliness checking must be carried out before the system is operational.

Certainly, the use of *active* components, replicated and distributed across the system is of great help to build fault-tolerant real-time systems. In this case, there must be a robust protocol to coordinate the replicas in a timely fashion,

C.A. Maziero et al. (Eds.): LADC 2005, LNCS 3747, pp. 154–173, 2005.

which means that extra computational efforts must be spent even in the absence of faults in order to make the system fault tolerant. An alternative approach is to have *passive* redundant components which can be activated upon error-detection. Since fault scenarios are exceptions, extra computational effort due to fault tolerance can be minimised. Although there is a higher time delay for detecting the error and recovering the system, the latter approach can be effective. Indeed, it is possible to introduce a greater level of flexibility in the system since the redundant component, when activated, may carry out alternative actions to recover or compensate the system from the specific detected error. Also, several modern programming languages allow for error-detection and recovery mechanisms to be programmed at the application level (*e.g.* exception handler) [4] so that the application needs are taken into account. Most importantly, both active and passive redundancy approaches can be jointly employed and in fact they may be designed to complement each other. For example, a transient fault may cause an active component to produce incorrect results. This error can be detected and a passive component can be activated to undo or redo the faulty component actions or even to silence it (*e.g.* shutting it down).

When applying the passive redundancy approach in the context of hard real-time systems a problem to be solved is how to compute the extra time required to execute the component actions when errors are detected. This is the focus of this paper. More specifically, we structured the system as a set of tasks, some of which execute only when errors are detected. Our goals are to: (a) determine whether the execution of system tasks meet their deadlines in the presence of faults; and (b) increase the system fault resilience by providing an appropriate task scheduling mechanism. To do so, we have developed a scheduling mechanism that can be adjusted (by priority assignment) to increase the fault tolerance capacity of the system and have derived a set of equations that are used to verify the system schedulability.

Our approach is based on determining (off-line) which priorities can be assigned to tasks so that more errors can occur without compromising the specified deadlines. As priority represent the urgency of execution, a task that has to be recovered (preferably) should have its recovery actions executing with higher priority. Determining the task priorities off-line is attractive because it provides a simple on-line scheduling criterion (*i.e.* the scheduler, at run-time, needs only to choose the highest priority task to execute). Also, complex and perhaps timing-consume criteria to determine the (best) priority assignments can be used.

The approach described in this paper is an extension of our former work [15], where a less restricted fault model is assumed. Indeed, here we assume that errors may take place at any time instead of considering that errors take place periodically in worst case.

The remainder of this paper is structured as follows. A brief literature review is given in the next section. The computation model is described in Section 3. Then, Section 4 presents some basic concepts on hard real-time scheduling and illustrates the addressed problem. Section 5 derives the schedulability analysis. The problem of searching for a priority assignment that improves the fault resilence of the system is addressed in Section 6. Some simulation results are also presented in this section. Then, in Section 7, our final comments are given.

2   Related Work

We identify two branches of work on providing fault tolerance in hard real-time systems, distributed protocols and scheduling. The former deals with coordinating the computation between different nodes when a distributed architecture is necessary [21,20,10]. On the other hand, for hard real-time systems, one has also to be concerned with the (local) computation carried out in the system nodes. Indeed, independently of whether the system is distributed or not, scheduling is a fundamental problem for real-time systems. In this section we sumarise only approaches focused on scheduling.

Scheduling for fault tolerance can further be divided into two categories: those approaches that take into account task replicas running in different nodes and those that focus on the execution of tasks considering only local computation. There are several examples of the former approach [11,2,9,10,16,17]. In this paper we deal with scheduling from the point of view of local nodes with the goal of providing fault tolerance within each node (if the system is distributed). Approaches to doing so usually consider two types of tasks running in the nodes, *primary* and *alternative*. Primary tasks represent the usual computation that needs to be performed in error-free scenarios. Alternative tasks contain actions that must be executed when some error is detected.

One of the first such a mechanism to schedule primary and alternative tasks was described by Liestman *et al.* [13]. This mechanism only deals with periodic tasks, whose periods have to be multiples of each other. The approach presented by Ghosh *et al.* [7] limits the recovery of faulty tasks to re-executing them. Only transient faults can be tolerated (*e.g.* design faults are not considered). An interesting approach to tolerating transient faults which is independent of the schedulability analysis being used has been described by Ghosh *et al.* [6]. However, only the re-execution of faulty tasks as a means of fault tolerance is assumed. Kandasamy *et al.* [8] describe a recovery technique that tolerates transient faults in an *off-line* scheduled distributed system. It is based on taking advantage of task set spare capacity. The amount of spare capacity is distributed over a given period so that task faults can be handled. Although tasks are assumed to be preemptive and their precedence relations are taken into account, only periodic tasks, whose periods are equal to deadlines, are considered.

Recently, an EDF (Earlier Deadline First) based scheduling, which takes the effects of transient faults into account, has been proposed [12]. Its basic idea is to simulate the EDF scheduler and to use slack time for executing task recoveries given a *fault pattern*. Fault patterns, which are the assumed maximum numbers of errors per task, must be known *a priori*. Task recoveries can be modelled as alternative tasks that are released after error-detection. Another EDF based scheduling approach for supporting fault-tolerant systems has been proposed by Caccamo *et al.* [5]. Their task model consists of instance skippable and fault-tolerant tasks. The former may allow the system to skip one instance once in a while. The latter is not skippable (*i.e.* all instances have to execute by their deadlines) and is composed of a primary and an alternative part. The primary part is scheduled on-line and provides high-quality service while the alternative one is scheduled off-line and provides acceptable services.

The approach presented by Ramos-Thuel *et al.* [19] is based on the *transient server* concept. Its basic idea is to explore the spare capacity of the task set to determine the maximum server capacity at each priority level. A server is an *a priori* created task used to service aperiodic requests. In their approach such requests are the detection of errors. The spare capacity allocated to the server is used for *on-line* dispatching decisions in the case of error occurrences. Although this approach seems interesting since higher priority levels are used to execute alternative tasks, a reasonable way of determining the server periods has not been presented.

A flexible approach that makes use of fixed-priority scheduling and response time analysis has been proposed by Burns *et al.* [3] and Punnekkat [18]. No restriction on alternative tasks is assumed. This approach shows that response time analysis can be straightforwardly adapted to take the execution of alternative tasks into account. Making use of this results, we have recently showed that non-standard priority assignments can be used to increase the fault resilience of the system [15]. Like Burns *et al.* we have restricted the fault model by assuming that there is a minimum time between consecutive errors.

In this paper we extend our former work [15] by removing this restriction on time between consecutive errors. By doing so, we take into consideration more general situations where errors may affect the execution of tasks at any time.

3   Computation Model

We assume that there is a set $\Gamma = \{\tau_1, \ldots, \tau_n\}$ of n tasks, called *primary tasks*, that must be scheduled by the system in the absence of errors. Any primary task τ_i in Γ has a period, T_i, a deadline D_i ($D_i \leq T_i$), and a worst-case computation time, C_i. Tasks can be periodic or sporadic. For sporadic tasks the period means the minimum inter-arrival time. Each primary task τ_i can have some *alternative tasks* associated with it. Each alternative task corresponds to a given action taken to recover τ_i from a given error. Any alternative task has a worst-case computation time, also called worst-case recovery time. For the sake of simplicity we denote $\overline{\tau}_i$ as the alternative task of τ_i whose worst-case recovery time is the largest one. Also, we assume that all alternative tasks associated with τ_i run at the same priority level. Hence, hereafter we do not include the details of individual alternative task per primary in the description we present. We only need to refer to $\overline{\tau}_i$ as the worst-case alternative task in case of errors in task τ_i.

Primary tasks are scheduled according to some fixed priority assignment algorithm, which attributes a distinct priority to each task τ_i in Γ. We consider n different priority levels $(1, 2, \ldots, n)$, where 1 is the lowest priority level. The alternative tasks of τ_i are assumed to execute at priority levels greater than or equal to τ_i's priority. We denote the priority of τ_i and $\overline{\tau}_i$ as $\mathsf{pr}(\tau_i)$ and $\mathsf{pr}(\overline{\tau}_i)$, respectively. When a primary task, say τ_i, and an alternative task, say $\overline{\tau}_j$, are ready to execute at the same priority level, we assume that $\overline{\tau}_j$ is scheduled first.

Alternative tasks represent some extra processing that is necessary to recover a task from a given erroneous state caused by a fault. Errors are detected at the task level. When an error interrupts the execution of a task, the system must schedule an appropriate alternative task, which is responsible for carrying out the error-processing procedure and has to finish by the deadline of its primary task. If other errors take place in the alternative tasks, we assume that it is scheduled again for re-execution. We also assume that there is no cost associated with any scheduling of primary or alternative tasks. These costs are assumed to be taken into account by the value of C_i and $\overline{C}_i$, respectively. Further, we assume that all errors are detected by the system and there is no fault propagation in the value domain (*i.e.* faults affect only the results produced by the executing task).

The kinds of fault with which we are dealing are those that can be treated at the task level. Consider for example *design* faults. It may be possible to use techniques such as *exception handling* or *recovery blocks* to perform appropriate recovery actions [3], modelled here as alternative tasks. In addition, one may consider some kind of *transient faults*, where either the re-execution of the faulty task or the execution of some compensation action is effective. For example, suppose that transient faults in a sensor (or network) prevent an expected signal from being correctly received (or received at all) by the control system. This kind of system fault can easily be modelled by alternative tasks, which can be released to carry out a compensation action. However, it is important to emphasise that we are not considering more severe kinds of fault that cannot be treated at the task level. For example, if a memory fault causes the value of one bit to be arbitrarily changed, the operating system may fail, compromising the whole system. Tolerating these kinds of fault requires spatial redundancy (perhaps using a distributed architecture) and is not covered in this paper. Our work fits the engineering approach that uses temporal redundancy at the processor level and spacial redundancy at the system level.

We derive the schedulability analysis of the system as a function of a fault resilience metric, denoted by N_E. The value of N_E represents the assumed maximum number of errors that task set may suffer. The goal of the analysis is to show: (a) whether or not the system is schedulable for a given value of N_E; and (b) whether or not a more resilient system can be built by assigning priorities to (alternative) tasks appropriately.

4   Initial Concepts

The schedulability derived in this paper is based on the well known response time analysis [1]. The basic idea is to compute the worst-case response time of each task in the system, R_i, and compare with its deadline. If $R_i \leq D_i$ for all tasks, then the system is schedulable. In the next two subsections we introduce some basic concepts in response time analysis and show how the effects of faults can be easily incorporated into the analysis. At the end of the section we illustrate potential advantages of having some alternative tasks running at higher priority levels.

4.1   Fault Free Scenarios

To compute R_i one has to consider the worst-case scenario. This happens when all other higher priority tasks τ_j are released at the same time as τ_i and the execution of τ_i and all higher priority tasks τ_j take C_i and C_j to complete, respectively. In this scenario, the value of R_i is given by C_i plus the sum all $n_j C_j$, where n_j is the maximum number of instance of τ_j that can occur during the execution of τ_i. Note that n_j depends on the response time of τ_i, i.e. $n_j = \lceil \frac{R_i}{T_j} \rceil$. Thus, R_i can be written as [1]:

$$R_i = C_i + \sum_{j \in hp(i)} \left\lceil \frac{R_i}{T_j} \right\rceil C_j, \tag{1}$$

where $hp(i)$ is the set of tasks that have higher priority than τ_i and $\lceil x \rceil$ returns the smallest integer that is greater than or equal to x.

Since the term R_i appears in both sides of equation (1), it is solved iteratively applying the relation given by equation (2) [1]. The iteration can start with $r_i^0 = C_i$, where r_i^k is the kth approximation to the true value of R_i. The interactions can be halted when $r_i^{k+1} > D_i$ or earlier if $r_i^{k+1} = r_i^k$. In the former case, the task is not schedulable, while the latter means that $R_i = r_i^k$.

$$r_i^{n+1} = C_i + \sum_{j \in hp(i)} \left\lceil \frac{r_i^n}{T_j} \right\rceil C_j. \tag{2}$$

4.2   Fault Scenarios

If an error occurs, from the point of view of τ_i, the worst case is when this error interrupts the execution the task with which is associated the longest alternative task among all tasks τ_k that can cause the interference in the execution of τ_i (including τ_i itself). Considering N_E errors and the fact that alternative tasks run with the same priority as their primary tasks, equation (1) can be extended as a function of N_E:

$$R_i(N_E) = C_i + \sum_{\tau_j \in hp(i)} \left\lceil \frac{R_i}{T_j} \right\rceil C_j + N_E \max_{\tau_k \in hpe(i)} \overline{C}_k, \tag{3}$$

where $hpe(i)$ is the set of tasks that have priorities higher of equal to the priority of τ_i. Indeed, for each error occurrence $\overline{C}_k$ time units are incorporated into the computation of R_i.

As an illustration, consider a set of 3 tasks and their alternative tasks shown in Table 1. The values in its last column are the worst-case response times for $N_E = 1$. To illustrate this, the following is what the iterative computation of R_3 looks like:

Table 1. A task set and the derived worst-case response times

| Task set | | | | | $N_E = 1$ |
Task T_i	C_i	$\overline{C}_i$	D_i	$pr(\tau_i)$	$R_i(1)$
τ_1 13	2	2	13	3	4
τ_2 25	3	3	25	2	8
τ_3 30	5	5	30	1	22

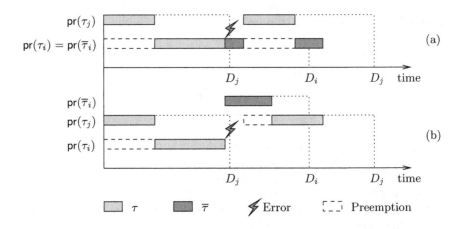

Fig. 1. Priority assignment in fault-scenarios

$$r_3^0(1) = 5$$

$$r_3^1(1) = 5 + \left\lceil \frac{5}{13} \right\rceil 2 + \left\lceil \frac{5}{25} \right\rceil 3 + 5 = 15$$

$$r_3^2(1) = 5 + \left\lceil \frac{15}{13} \right\rceil 2 + \left\lceil \frac{15}{25} \right\rceil 3 + 5 = 22$$

$$r_3^3(1) = 5 + \left\lceil \frac{22}{13} \right\rceil 2 + \left\lceil \frac{22}{25} \right\rceil 3 + 5 = 22$$

$$R_3(1) = 22$$

The alternative tasks of Table 1 inherit the priorities of their primary tasks, which are given by some conventional priority assignment (*e.g.* DM - Deadline Monotonic). However, when errors take place, it may be advantageous to execute alternative tasks at higher priority levels. The intuition is that, after an error, tasks (or their recovery/compensation actions) certainly have a shorter period of time to meet their deadlines.

Consider Fig. 1, where the execution time line of two tasks, $\{\tau_i, \tau_j\}$, is shown. The priorities of τ_j and τ_i are assigned by the deadline monotonic approach, *i.e.* $pr(\tau_j) > pr(\tau_i)$, since $D_j < D_i$. An error interrupts τ_i just before the end of its execution, as illustrated. In scenario (a), it is assumed that $\overline{\tau}_i$ is executed with priority level $pr(\tau_i)$ whereas in scenario (b) the priority of $\overline{\tau}_i$ is greater than the priority of τ_j. As can be seen, in Fig. 1(a) τ_i is not schedulable due to the preemption caused by the execution of the second release of τ_j. This preemption is avoided by executing $\overline{\tau}_i$ at a higher priority level, as illustrated in Fig. 1(b). In the next section, a new set of equations is derived to take into consideration this kind of priority assignment.

5 Schedulability Analysis: A More General Approach

Let Γ be a given task set and $N_E \geq 0$ an integer. The main goal of this section is to develop schedulability analysis to check whether or not Γ is schedulable if up to N_E errors take place during the execution of any task in Γ. Unlike the previous section, we assume here that alternative tasks may have higher priorities than their respective primary tasks. This assumption, as will be seen, makes the analysis much more complex. The problem of determining the priorities of the alternative tasks is postponed to Section 6. Here we focus only on the analysis by assuming that the priority assignment is known.

The strategy to derive the analysis is the following. First, in Section 5.2, we consider that τ_i is fault-free. In this case we say that there are only *external* errors regarding τ_i (*i.e.* errors take place in other tasks but τ_i). Thus, any task (primary or alternative) that has priorities higher than or equal to the priority of τ_i may interfere in its execution. Then, we consider the case of some error in τ_i, *i.e.* there is some *internal* error. This case, addressed in Sections 5.3 and 5.4, is characterised by the fact that the execution of $\overline{\tau}_i$ may interfere in the execution of higher priority tasks. As will be seen, this case is more complicated because there are two phases during the response time of τ_i: before and after the release of $\overline{\tau}_i$. Because of this, the problem of finding what the distribution of errors leads to the worst-case scenario has to be considered. This problem is addressed in Section 5.5. Before describing the analysis, we present some definitions in the next section.

5.1 Definitions

A particular choice of priorities for alternative tasks, named *priority configuration*, is defined as follows:

Definition 1 (Priority configuration). *A priority configuration P_x is a tuple $\langle h_{x,1}, h_{x,2}, \ldots, h_{x,n} \rangle$, where $0 \leq h_{x,i} < i$ and $h_{x,i} = \mathsf{pr}(\overline{\tau}_i) - \mathsf{pr}(\tau_i)$.*

Note that $h_{x,i}$ represents the priority increment for task $\overline{\tau}_i$ in relation to the priority of its primary task τ_i. The definition of $h_{x,i}$ bounds the priority of $\overline{\tau}_i$ from τ_i's priority to the highest priority level. Lower priority levels are not considered. For example, consider $P_x = \langle 0, 0, \ldots, 0 \rangle$ a priority configuration. This means that any alternative task executes at the same priority level as the primary task with which it is associated. For $P_x = \langle 0, 0, \ldots, 0, 1 \rangle$, all tasks execute at their original priority level apart from $\overline{\tau}_n$, which executes one priority level above its primary task.

Given a priority configuration P_x, the following subsets of Γ regarding the priority of task $\tau_i \in \Gamma$ can be defined:

- $\mathsf{ip}(x, i)$. These are the tasks that may interfere in the response time of τ_i as regards priority configuration P_x if an error occurs. More formally, $\mathsf{ip}(x, i) = \{\tau_j \in \Gamma \mid h_{x,j} + \mathsf{pr}(\tau_j) \geq \mathsf{pr}(\tau_i)\}$.
- $\mathsf{sp}(x, i)$. Tasks that belong to such a subset do not suffer any extra interference when errors interrupt the execution of τ_i as regards priority configuration P_x. This is because their priorities are superior to $\mathsf{pr}(\overline{\tau}_i)$. More formally, $\mathsf{sp}(x, i) = \{\tau_j \in \Gamma \mid \mathsf{pr}(\tau_j) > h_{x,i} + \mathsf{pr}(\tau_i)\}$.
- $\mathsf{ipe}(x, i)$. This subset is defined as follows. If $h_{x,i} = 0$, then $\mathsf{ipe}(x, i) = \mathsf{ip}(x, i)$. Otherwise (when $h_{x,i} > 0$), $\mathsf{ipe}(x, i) = \mathsf{ip}(x, i) - \{\tau_i\}$. This subset is particularly useful for modelling cases where errors may interrupt task τ_i since the maximum interference its recovery suffers depends on whether or not $\mathsf{pr}(\tau_i) = \mathsf{pr}(\overline{\tau}_i)$. The meaning of this subset will be clearer later on when we describe the effects of internal errors.

5.2   External Errors

The computation of $R_i^{ext}(x, N_E)$, the worst-case response time of task τ_i due to external errors, is straightforward. This is because $\overline{\tau}_i$ does not need to be considered. In this situation, the worst-case scenario, as for task τ_i, can be described as follows: (a) every task that executes requires its worst-case execution time; (b) errors take place just before the end of the execution of tasks; (c) just before the release of τ_i some alternative task with maximum recovery time among all tasks in $\mathsf{ip}(x, i) - \{\tau_i\}$ is released; and (d) all tasks in $\mathsf{hp}(i)$ are released at the same time as τ_i. Therefore, one has to take into account the time to execute τ_i plus all tasks in $\mathsf{hp}(i)$ and the time to recover the faulty task times the maximum number of errors that may occur over $R_i^{ext}(x, N_E)$. This scenario yields equation (4):

$$R_i^{ext}(x, N_E) = C_i + \sum_{\tau_j \in \mathsf{hp}(i)} \left\lceil \frac{R_i^{ext}(x, N_E)}{T_j} \right\rceil C_j + N_E \max_{\tau_k \in \mathsf{ip}(x,i) - \{\tau_i\}} (\overline{C}_k), \quad (4)$$

It is not difficult to see that if $\overline{C}_i < \max_{\tau_k \in \mathsf{ip}(x,i)}(\overline{C}_k)$, equation (4) gives the worst-case response time of τ_i. Indeed, if only external errors take place, the meaning of equation (4) is similar to equation (3). Also, it is not difficult to see that if some internal error happens, $\overline{C}_i$ cannot cause higher interference than task $\overline{\tau}_k$. Thus, the lemma bellow follows [14]:

Lemma 1. *Let Γ be a fixed-priority set of primary tasks and their respective alternative tasks. Suppose that Γ is subject to faults so that there are at most $N_E \geq 0$ errors during the execution of any task. Also, let P_x be a priority configuration for the alternative tasks. If $\overline{C}_i < \max_{\tau_k \in \mathsf{ip}(x,i)}(\overline{C}_k)$, $R_i^{ext}(x, N_E)$ represents the worst-case response time of τ_i regardless of whether or not the execution of τ_i is interrupted by some error.*

5.3   Internal Errors: Some Intuition

Note that the input parameter N_E does not say much about how the error occurrences are distributed. For example, assume that $N_E = 2$. In this case there are two scenarios that must be considered regarding task $\tau_i \in \Gamma$: (a) an error interrupts the execution of some other task τ_j before the first internal error hits τ_i; or (b) the second error takes place after τ_i suffers the internal error. The problem is that it is not possible to know beforehand which of these scenarios represent the worst-case. Thus, it is convenient to define N_E as follows. For each task $\tau_i \in \Gamma$:

$$N_E = N_i^0 + N_i^1 \tag{5}$$

The terms N_i^0 and N_i^1 stand for the maximum number of errors that may take place before and after (or at) the time the first internal error hits τ_i, respectively. Hence, if $N_E = 2$, the possible combinations are: $N_i^0 = 1$ and $N_i^1 = 1$ for scenario (a); and $N_i^0 = 0$ and $N_i^1 = 2$ for scenario (b). The combination $N_i^0 = 2$ and $N_i^1 = 0$ does not need to be considered since it would imply that all N_E errors are external. Therefore, in general, if $N_E = k$, there are k different scenarios that must be analysed in order to determine which one represents the worst-case.

The worst-case response time of a task τ_i considering the occurrence of some internal error is now a function of P_x, N_i^0 and N_i^1 and is denoted by $R_i^{int}(x, N_i^0, N_i^1)$. The approach to computing its value is divided into two steps. This is because the procedure to calculate $R_i^{int}(x, N_i^0, N_i^1)$ has to take into account two levels of priorities (before and after the first internal error) when $\mathsf{pr}(\overline{\tau}_i) > \mathsf{pr}(\tau_i)$.

It is important to emphasise that the values of both N_i^0 and N_i^1 must be set such that they lead to the worst-case value of $R_i^{int}(x, N_i^0, N_i^1)$. A simple procedure for determining the appropriate values of N_i^0 and N_i^1 is iterative and must evaluate all scenarios. This procedure is explained shortly after the descriptions of the equations that give $R_i^{int}(x, N_i^0, N_i^1)$.

5.4   Internal Errors: The Derivation

Assume that the values of N_i^0 and N_i^1 are known and that at least an internal error takes place at some time t (see Fig. 2). What has to be computed is the maximum time $\overline{\tau}_i$ lasts if it is subject to both other possible errors and the interference due to tasks in $\mathsf{sp}(x, i)$ from t onwards. This time is represented in the figure by $R_i^{int^1}(x, N_i^1)$.

In the worst case there may be N_i^1 errors over the period $R_i^{int^1}(x, N_i^1)$. The first error accounts for $\overline{C}_i$, while the others may cause the release of the recovery of any task in $\mathsf{sp}(x, i) \cup \{\tau_i\}$. The worst case is when all $N_i^1 - 1$ other errors interrupt a task in $\mathsf{sp}(x, i) \cup \{\tau_i\}$ that has the longest recovery time.[1] Therefore, the value of $R_i^{int^1}(x, N_E^1)$ can be computed iteratively by

---

[1] Note that here a generic situation is assumed. However, in practice one can consider that all errors from t onwards are internal, due to lemma 1.

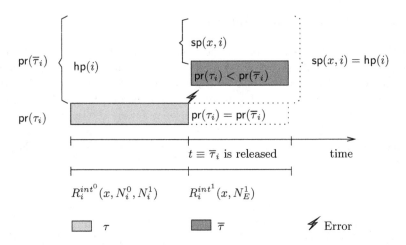

Fig. 2. Illustration of the derivation of R_i^{int}

$$R_i^{int^1}(x, N_i^1) = \overline{C}_i + \sum_{\tau_j \in \mathsf{sp}(i)} \left\lceil \frac{R_i^{int^1}(x, N_i^1)}{T_j} \right\rceil C_j + \left(N_i^1 - 1\right) \max_{\tau_k \in \mathsf{sp}(x,i) \cup \{\tau_i\}} (\overline{C}_k).$$
(6)

The computation of $R_i^{int^0}(x, N_i^0, N_i^1)$ is slightly more complex. Let us analyse it considering two cases depending on the values of $\mathsf{pr}(\tau_i)$ and $\mathsf{pr}(\overline{\tau}_i)$:

When $\mathsf{pr}(\tau_i) < \mathsf{pr}(\overline{\tau}_i)$. This means that $\overline{\tau}_i$ executes at a higher priority level. Note that, in this case, knowing $R_i^{int^0}(x, N_i^0, N_i^1)$ is equivalent to knowing the relative earliest possible release time of τ_i so that it suffered the first internal error at time t, as illustrated in Fig. 2.

During $R_i^{int^0}(x, N_i^0, N_i^1)$, τ_i may suffer the preemption of tasks in $\mathsf{hp}(i)$ and possibly the recoveries of tasks in $\mathsf{ip}(i) - \{\tau_i\}$ due to other errors. It is important to note that τ_i has to be removed from the set of tasks that may suffer errors in this phase because, by assumption, the first internal error occurs at time t. Indeed, if there was an earlier internal error, then $\overline{\tau}_i$ would be released earlier and so it would finish earlier. It is clear that this situation does not represent the worst-case scenario.

When $\mathsf{pr}(\tau_i) = \mathsf{pr}(\overline{\tau}_i)$. Unlike the former case, the maximum interference during the period $R_i^{int^0}(x, N_i^0, N_i^1)$ can take place when all errors are internal since both τ_i and its alternative task run at the same priority level. This situation happens, for example, when $\overline{C}_i = \max_{\tau_k \in \mathsf{ip}(x,i)}(\overline{C}_k)$. As a result, instead of considering errors in $\mathsf{ip}(x,i) - \{\tau_i\}$, one should consider errors in the whole $\mathsf{ip}(x,i)$.

In summary, as far as possible errors during $R_i^{int^0}(x, N_i^0, N_i^1)$ are concerned, when $\mathsf{pr}(\tau_i) = \mathsf{pr}(\overline{\tau}_i)$, one has to consider errors in $\mathsf{ip}(x,i) - \{\tau_i\}$. Otherwise, errors in $\mathsf{ip}(x,i)$ should be taken into account. This is the main difference between the cases analysed above. In order to join both cases together in a single equation, the set $\mathsf{ipe}(x, i)$ can be used. Indeed, errors during the interval $R_i^{int^0}(x, N_i^0, N_i^1)$ may take place in any task in $\mathsf{ipe}(x, i)$.

The equation that gives $R_i^{int^0}(x, N_i^0, N_i^1)$ can now be derived. It has to take into account: the worst-case execution time of τ_i (C_i); the interference due to tasks in $\mathsf{hp}(i)$; and possible recoveries of tasks in $\mathsf{ipe}(x, i)$. Note that some releases of tasks in $\mathsf{sp}(x, i)$ may already have been taken into account when computing $R_i^{int^1}(x, N_i^1)$. This means that one has to take care not to include the same task in $\mathsf{sp}(x, i)$ twice. In other words, one has to subtract for each task in $\mathsf{sp}(x, i)$ and each error occurrence the interference already computed in $R_i^{int^1}(x, T_E)$.

From the description above, equation (7) gives the value of $R_i^{int^0}(x, N_i^0, N_i^1)$. Note that instead of computing the worst-case interference due to tasks in $\mathsf{hp}(i)$, this computation is split as for two complementary subsets, $\mathsf{hp}(i) - \mathsf{sp}(x, i)$ and $\mathsf{sp}(x, i)$. This is to avoid the computation of tasks in $\mathsf{sp}(x, i)$ more than once, as previously mentioned. This is done by subtracting $\lceil \frac{R_i^{int^1}(x, N_i^1)}{T_l} \rceil \overline{C}_l$ for each task $\tau_l \in \mathsf{sp}(x, i)$.

$$R_i^{int^0}(x, N_i^0, N_i^1) = C_i + \sum_{\tau_j \in \mathsf{hp}(i) - \mathsf{sp}(x,i)} \left\lceil \frac{R_i^{int^0}(x, N_i^0, N_i^1)}{T_j} \right\rceil C_j +$$
$$\sum_{\tau_l \in \mathsf{sp}(x,i)} \left(\left\lceil \frac{R_i^{int}(x, N_i^0, N_i^1)}{T_l} \right\rceil - \left\lceil \frac{R_i^{int^1}(x, N_i^1)}{T_l} \right\rceil \right) C_l +$$
$$N_i^0 \max_{\tau_k \in \mathsf{ipe}(x,i)} (\overline{C}_k). \tag{7}$$

The final value of $R_i^{int}(x, N_i^0, N_i^1)$ is given by summing up $R_i^{int^0}(x, N_i^0, N_i^1)$ and $R_i^{int^1}(x, N_i^1)$:

$$R_i^{int}(x, N_i^0, N_i^1) = R_i^{int^0}(x, N_i^0, N_i^1) + R_i^{int^1}(x, N_i^1). \tag{8}$$

5.5   Number of Errors

The problem of finding out appropriate values of N_i^0 and N_i^1 is solved iteratively. The idea is to use the schedulability analysis to check which combination of N_i^0 and N_i^1 leads to the worst-case scenario. The algorithm to do so is described in Fig. 3.

The idea of the algorithm is to *distribute* N_E error occurrences during the worst-case response time of a task τ_i. One error hits task τ_i (by assumption). The other $N_E - 1$ errors are considered to be either in $R_i^{int^0}(x, N_i^0, N_i^1)$ or in $R_i^{int^1}(x, N_E)$, depending on which choice gives higher values for $R_i^{int}(x, N_i^0, N_i^1)$. If the task set is unschedulable in some iteration, then the task set does not tolerate N_E errors. Otherwise, the final values of N_i^0 and N_i^1 are given by N^0 and N^1, respectively. At the end of the algorithm, the value of variable R contains the worst-case response time considering internal errors as long as the task set is schedulable in P_x with N_E errors (some of them internal).

// This algorithm needs to be executed for each task $\tau_i \in \Gamma$ that has:
// (a) $R_i^{ext}(x, N_E) \leq D_i$; (b) $\text{pr}(\overline{\tau}_i) > \text{pr}(\tau_i)$; and (c) $\overline{C}_i > \max_{\tau_k \in \text{ip}(x,i) - \{\tau_i\}}(\overline{C}_k)$.

(1) $N^0 \leftarrow 0; N^1 \leftarrow 1; k \leftarrow 1$
(2) $R \leftarrow R_i^{int}(x, N^0, N^1)$
(3)   **while** $k < N_E \wedge \Gamma$ is schedulable **do**
(4) $R^0 \leftarrow R_i^{int}(x, N^0 + 1, N^1)$
(5) $R^1 \leftarrow R_i^{int}(x, N^0, N^1 + 1)$
(6)   **if** $(R^0 > R^1)$ **then**
(7) $R \leftarrow R^0$
(8) $N^0 \leftarrow N^0 + 1$
(9)   **else**
(10) $R \leftarrow R^1$
(11) $N^1 \leftarrow N^1 + 1$
(12)   **endif**
(13) $k \leftarrow k + 1$
(14)   **endwhile**
(15)   **if** Γ is schedulable **then**
(16)   // R is the solution for $R_i^{int}(x, N_i^1, N_i^1)$
(17)   // where $N_i^1 = N^0$ and $N_i^1 = N^1$
(18)   **else**
(19)   // Task set cannot cope with N_E errors
(20)   **endif**

Fig. 3. Procedure to determine the values of N_i^0 and N_i^1

Initially, $N^0 = 0$ and $N^1 = 1$. The initial value of N^1 accounts for the first assumed internal error. Then, in each iteration, either N^0 or N^1 is increased by 1 depending on which one makes the value of $R_i^{int}(x, N^0, N^1)$ bigger. The strategy of increasing the number of errors by one at each time is for the sake of performance. Indeed, equations (6) and (7) are monotonically non-decreasing in function of the number of errors and they are also solved iteratively. Hence, the initial value to solve them for a particular choice of N^0 and N^1 can be the solutions obtained in the previous iteration. For example, let the values of $R_i^{int}(x, 1, 1)$ and $R_i^{int}(x, 0, 2)$ be calculated in iteration it, say. If the task set is schedulable and it $< N_E$, either $R_i^{int}(x, N^0 + 1, N^1)$ or $R_i^{int}(x, N^0, N^1 + 1)$ will be computed in the $(it + 1)^{th}$ iteration. If so, such a computation can start from the previously computed values in iteration it (variable R in the algorithm). The implementation details to do so are not explicitly expressed in the algorithm of Fig. 3 but it can easily be carried out by integrating the algorithm with the iterative procedure that solves equations (6) and (7).

Clearly, the computational effort to calculate $R_i^{int}(x, N_i^0, N_i^1)$ is higher when compared to the computation of $R_i^{ext}(x, N_E)$. Therefore, it is important to observe some aspects related to the need for performing the algorithm of Fig. 3. Indeed, one only needs to calculate $R_i^{int}(x, N_i^1, N_i^1)$ if the following conditions hold:

Table 2. An illustrative task set and the values of worst-case response times (in bold)

Task set	$N_E = 2$				
	$\langle 0,0,2 \rangle$				
Task T_i	C_i	$\overline{C}_i$	D_i	R_i^{int}	R_i^{ext}
τ_1 13	2	2	13	–	**12**
τ_2 25	3	4	25	–	**17**
τ_3 30	5	5	30	**21**	20

(a) $\forall \tau_i \in \Gamma$: $R_i^{ext}(x, N_E) \le D_i$; (b) $\mathsf{pr}(\overline{\tau}_i) > \mathsf{pr}(\tau_i)$; and (c) $\overline{C}_i > \max_{\tau_k \in \mathsf{ip}(x,i) - \{\tau_i\}}$ $(\overline{C}_k)$. If condition (a) or (c) does not hold, the computation of $R_i^{int}(x, N_i^0, N_i^1)$ is irrelevant because either the task set is already unschedulable or by lemma 1 it is known that $R_i(x, N_E) = R_i^{ext}(x, N_E)$. Moreover, should $\mathsf{pr}(\overline{\tau}_i)$ equal $\mathsf{pr}(\tau_i)$, any solution of equation (5) can be used. For example, $N_i^1 = 0$ and $N_i^1 = N_E$. This is because under this condition $\mathsf{ipe}(x, i) = \mathsf{ip}(x, i)$.

Should the values of N_i^0 and N_i^1 be determined by the algorithm of Fig. 3 for some task $\tau_i \in \Gamma$, its worst-case response time is given by

$$R_i(x, N_i^0 + N_i^1) = \max \left[R_i^{ext}(x, N_i^0 + N_i^1), R_i^{int}(x, N_i^0, N_i^1) \right] . \qquad (9)$$

Otherwise, it is given simply by taking $R_i(x, N_E) = R_i^{ext}(x, N_E)$.

5.6   An Illustrative Example

Consider the task set in Table 2. This task set, with three tasks, is the same as that given in Table 1 but with $\overline{C}_2 = 4$ time units. Assume that the priority of primary tasks are given according to DM and let $P_x \langle 0, 0, 2 \rangle$ and $N_E = 2$. The values of $R_i^{ext}(x, 2)$ for each of the three tasks are iteratively calculated by equation (4), similarly to the explanation given in Section 4. The found values are 12, 17 and 20, as indicated in the table.

As can be seen, considering only external errors, the task set is schedulable. However, as $\overline{C}_3 > \max_{\tau_k \in \mathsf{ip}(x,3) - \{\tau_3\}} (\overline{C}_k)$, $R_3^{int}(x, N_3^0, N_3^1)$ needs to be computed since $R_3^{ext}(x, N_E)$ may not give the worst-case response time. In order to do so, the algorithm of Fig. 3 is performed. Firstly, the values of $R_i^{int^1}(x, 1)$ and $R_i^{int^0}(x, 0, 1)$ are calculated (observe that $\mathsf{sp}(x, 3) = \emptyset$). We have that $r_3^{int^{1^0}}(x, 1) = r_3^{int^{1^1}}(x, 1) = 5$ and so $R_3^{int^1}(x, 1) = 5$. The value of $R_3^{int^0}(x, 0, 1)$ equals 10:

$$r_3^{int^{0^0}}(x, 0, 1) = 5 \,,$$

$$r_3^{int^{0^1}}(x, 0, 1) = 5 + \left\lceil \frac{5}{13} \right\rceil 2 + \left\lceil \frac{5}{25} \right\rceil 3 = 10 \,,$$

$$r_3^{int^{0^2}}(x, 0, 1) = 5 + \left\lceil \frac{10}{13} \right\rceil 2 + \left\lceil \frac{10}{25} \right\rceil 3 = 10 \,.$$

Consequently, $R_3^{int}(x, 0, 1) = 15$. Then, the iterative procedure starts (lines 3-14 of Fig. 3), where the values of both $R_3^{int}(x, 1, 1)$ and $R_3^{int}(x, 0, 2)$ are computed.

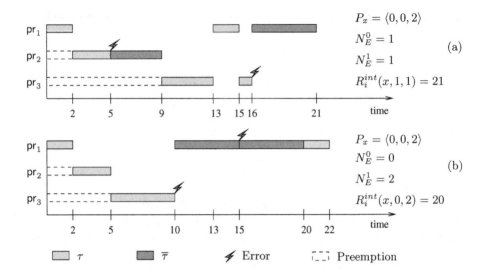

Fig. 4. Two possible fault scenarios for task τ_3 and $N_E = 2$

First, consider $R_3^{int}(x, 1, 1)$, which is obtained by equation (8), i.e. $R_3^{int^0}(x, 1, 1) + R_3^{int^1}(x, 1)$. $R_3^{int^1}(x, 1) = 5$ has been computed earlier. The computation of $R_3^{int^0}(x, 1, 1)$ is as follows. Since $R_3^{int^0}(x, 0, 1) = 10$, $r_3^{int^{0^0}}(x, 1, 1) = 10$. Carrying out the iterative procedure, one can see that $R_3^{int^0}(x, 1, 1) = 16$ since

$$r_3^{int^{0^2}}(x, 1, 1) = r_3^{int^{0^3}}(x, 1, 1) = 5 + \left\lceil \frac{16}{13} \right\rceil 2 + \left\lceil \frac{16}{25} \right\rceil 3 + 1 \times 4 = 16.$$

Thus, $R_3^{int}(x, 1, 1) = R_3^{int^0}(x, 1, 1) + R_3^{int^1}(x, 1) = 21$. Then, $R_3^{int}(x, 0, 2)$ is computed. The value of $R_3^{int^1}(x, 2)$ is equal to 10 since $r_3^{int^{1^0}}(x, 2) = r_3^{int^{1^1}}(x, 2) = 5 + (2-1)5 = 10$. Also, starting from $r_3^{int^{0^0}}(x, 0, 2) = 10$ since $R_3^{int^0}(x, 0, 1) = 10$, one will find that $R_3^{int^0}(x, 0, 2) = 10$. Hence, $R_3^{int}(x, 0, 2) = R_3^{int^0}(x, 0, 2) + R_3^{int^1}(x, 2) = 20$.

The algorithm stops after the first iteration since $N_E = 2$. As can be seen, the worst-case scenario for τ_3 with $P_x = \langle 0, 0, 2 \rangle$ and $N_E = 2$ is when one error hits τ_2 before the internal error in τ_3. This is because when $N_i^0 = 1$, τ_3 suffers interference from an extra release of τ_1. Fig. 4 illustrates this behaviour, where the two scenarios are presented. Scenario (a) represents the worst case and scenario (b) is when both errors are internal.

6   Priority Assignment and Evaluation

Once one can find the worst-case response time of each task given N_E errors, there is another problem to be solved: determining which priority configuration

P_x leads to the maximum value of N_E without making any task missing its deadline. Our approach to solving this problem is iterative. For the sake of space, only the general idea of the priority configuration search algorithm is presented here. The algorithm is very similar to the one previously published [15]. More details can be found in this or in other publication [14].

6.1   Searching for the Priority Configuration

Let $N_e(x)$ denote the maximum N_E that the task set can cope with in priority configuration P_x. $N_e(x)$ can be implemented as a binary search using the equations previously described in Section 5. The interval of the search can be set to $[0, \min(1, \lfloor \frac{D_i - C_i}{\overline{C}_i} \rfloor)]$, for example. Clearly, no task τ_i can cope with more than $\frac{D_i - C_i}{\overline{C}_i}$ (in the worst case). Note that by definition, no task set is schedulable in P_x if $N_e(x) + 1$ errors take place. The sketch of the algorithm is as follows:

1. Initially, let $P_x = \langle 0, 0, \ldots, 0 \rangle$ and let $N_E = N_e(x)$.
2. Repeat
    (a) Compute $R_i(x, N_E)$, $i = 1, \ldots, n$.
    (b) If the task set is schedulable, make $N_E = N_E + 1$. Save the value of P_x.
    (c) Otherwise
        i. If there is some unschedulable task due to external errors, stop searching priority configuration.
        ii. Look for the lowest priority task $\tau_j \in \mathsf{sp}(x, i)$ so that

$$\left\lceil \frac{R_i^{int}(x, N_i^0, N_i^1)}{T_j} \right\rceil > \left\lceil \frac{R_i^{int^0}(x, N_i^0, N_i^1)}{T_j} \right\rceil . \tag{10}$$

        iii. If there is such τ_j, make $\mathsf{pr}(\overline{\tau}_i) = \mathsf{pr}(\tau_j)$ and $N_E = \max(N_E, N_e(x))$. Otherwise stop searching priority configuration.
3. The last saved priority configuration is returned.

Once $N_e(x)$ is determined, the goal of the algorithm is to find out another priority configuration by raising the priority of some alternative task so that the task set is schedulable with $N_E = N_e(x) + 1$. To do so, one needs to decrease the values of $R_i^{int}(x, N_e(x) + 1)$ by increasing priorities of their alternative tasks if $R_i^{int}(x, N_e(x) + 1) > D_i$. However, these priorities only need to be raised if it is possible to decrease the number of preemption of higher priority tasks τ_j without making them unschedulable. In other words, if there is a task $\tau_j \in \mathsf{sp}(x, i)$ which is executed over the period of time $R_i^{int^1}(x, N_i^0, N_i^1)$ - see condition (10) - then making $\mathsf{pr}(\overline{\tau}_i) = \mathsf{pr}(\tau_j)$ will decrease the worst-case response time of τ_i due to internal errors. Carrying out the described procedure until it is no longer possible to decrease the values of $R_i^{int}(x, N_e(x) + 1) > D_i$, one will find the best priority configuration for fault tolerance purposes [14,15].

6.2 Results of Experiments

Some experiment results are shown in Fig. 5. These results were collected from carrying out the approach for a large number of task sets. The task sets, with 10 tasks each, were randomly generated as follows. The values of worst-case computation time were generated according to an exponential distribution with mean $U/10$, where U is the processor utilisation. The periods and deadlines of tasks were assigned according to a uniform distribution with minimum and maximum values set to 50 and 5,000, respectively. Deadlines were allowed to

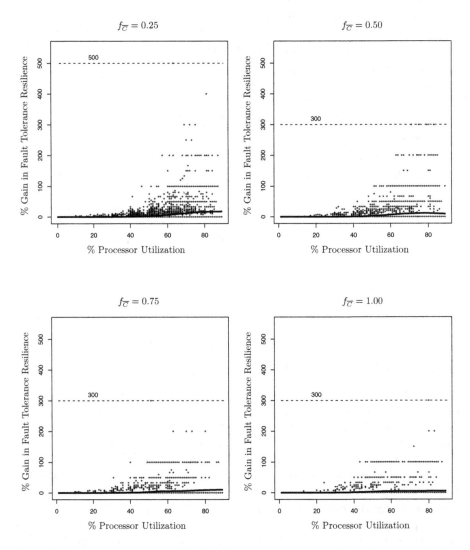

Fig. 5. Improvement in terms of fault resilience measured as obtained increase in N_E

be less than or equal to periods. We used the deadline monotonic algorithm to assign the priorities of primary tasks. We did not consider processor utilisation higher than 0.9 since it is difficult to guarantee the schedulability of the task set under error occurrences (*i.e.* most of the time it is not possible to tolerate even one fault at these higher processor utilisations).

Each one of the graphs represents the obtained gains in terms of increasing in N_E for 18,000 task sets. For each of the four graphs, a different value of recovery factor ($f_{\overline{C}}$) was used. The recovery factor is an input parameter of the simulation and was used to bound the worst-case execution time of the alternative tasks so that the values of $\overline{C}_i$ were generated according to a uniform distribution between 1 and $f_{\overline{C}}C_i$. For example, if $f_{\overline{C}} = 1$, $\overline{C}_i \leq C_i$.

As can be seen from Fig. 5, the gains obtained in terms of fault resilience as measured by N_E can be significant (up to 500% for $f_{\overline{C}} = .25$) and, as expected, is higher for lower values of $f_{\overline{C}}$. We can also observe that there were lower gains for lower processor utilisations. This can be explained by the fact that in these cases there is higher spare time available. This spare time can be used to carry out fault tolerance assuming higher values of N_E. Promoting the priority of alternative tasks for these cases, therefore, has lower impact in fault resilience since it is already high.

As illustration, consider Table 3, which shows the values of the worst-case response time due to external and internal errors of each task of a task set collected from the simulation. The values of the worst-case response time are in bold. It is worth emphasising the fact that in practice one does not need to perform algorithm 3 to compute the values of worst-case response time due to internal errors for all tasks. This is because of the reasons mentioned in Section 5. For example, for $P_x = \langle 0, 0, \ldots, 0 \rangle$ making $N_i^0 = 0$ and $N_i^1 = N_E$ for all tasks suffices. Also, for $P_x = \langle 0, 0, \ldots, 0, 9 \rangle$ the algorithm 3 only needs to be carried out with respect to τ_{10} (by lemma 1). Nevertheless, all values of $R_i^{int}(x, N_i^0, N_i^1)$ are shown in the table for the sake of illustration.

Table 3. Illustration of the improvement in fault tolerance resilience

Task set					$P_x = \langle 0,0,\ldots,0 \rangle$ $N_e(x) = 1$		$P_x = \langle 0,0,\ldots,0,9 \rangle$ $N_e(x) = 3$	
Task	T_i	C_i	$\overline{C}_i$	D_i	R_i^{int}	R_i^{ext}	R_i^{int}	R_i^{ext}
τ_1	4016	205	81	4011	**286**	205	448	**1303**
τ_2	4056	304	84	4031	**593**	590	761	**1607**
τ_3	4279	528	46	4034	1083	**1121**	1251	**2135**
τ_4	4363	99	88	4042	**1224**	1220	1400	**2234**
τ_5	4980	9	1	4061	1146	**1233**	1322	**2243**
τ_6	4164	17	2	4138	1164	**1250**	1340	**2260**
τ_7	4341	181	96	4197	**1439**	1431	1631	**2441**
τ_8	4518	90	49	4273	1482	**1529**	1674	**2531**
τ_9	4487	136	112	4305	**1681**	1665	1905	**2267**
τ_{10}	4643	1768	366	4490	**3703**	3449	**4435**	3673

7 Conclusion

In this paper we have presented an approach to increasing the fault tolerance resilience of hard real-time task sets in the context of fixed priority scheduling. The priorities of tasks are determined off-line so that the system can cope with more errors during their execution. To do so, a new framework to analyse the system under fault scenarios and an algorithm to search for the best priority configuration were derived. Both the analysis and priority configuration search algorithm were an extension of our former work [15], which assumed that there is a known minimum time interval between consecutive errors. Here this assumption was removed.

The proposed approach was extensively evaluated by simulation. Results from the experiments indicate that there are benefits in applying our approach. Indeed, for some cases significant gains (up to 500%) in terms of fault resilience was obtained for some cases.

The approach described in this paper takes into consideration the worst-case scenario to derive the priority assignments. It would be interesting to investigate ways of varying priorities dynamically in order to take advantage of spare capacities in the system. This will be part of our future work.

References

1. N. C. Audsley, A. Burns, M. Richardson, K. Tindell, and A. J. Wellings. "Applying New Scheduling Theory to Static Priority Pre-Emptive Scheduling". *Software Engineering Journal*, 8(5):284–292, 1993.
2. A. A. Bertossi, L. V. Mancini, and F. Rossini. "Fault-Tolerant Rate-Monotonic First-Fit Scheduling in Hard-Real-Time Systems". *IEEE Transaction on Parallel and Distributed Systems*, 10(9):934–945, 1999.
3. A. Burns, R. I. Davis, and S. Punnekkat. "Feasibility Analysis of Fault-Tolerant Real-Time Task Sets". In *Proc. of the Euromicro Real-Time Systems Workshop*, pages 29–33. IEEE Computer Society Press, 1996.
4. A. Burns and A. J. Wellings. *Real-Time Systems and Programming Languages*. Addison-Wesley, 3rd edition, 2001.
5. M. Caccamo and G. Buttazzo. "Optimal Scheduling for Fault-Tolerant and Firm Real-Time Systems". In *Proc. of the 5th Conference on Real-Time Computing and Applications (RTCSA)*, pages 223–231, 1998.
6. S. Ghosh, R. G. Melhem, and D. Mossé. "Enhancing Real-Time Schedules to Tolerate Transient Faults". In *Proc. of the 16th Real-Time Systems Symposium (RTSS)*, pages 120–129. IEEE Computer Society Press, 1995.
7. S. Ghosh, R. G. Melhem, D. Mossé, and J. S. Sarma. Fault-Tolerant Rate-Monotonic Scheduling. *Real-Time Systems*, 15(2):149–181, 1998.
8. N. Kandasamy, J. P. Hayes, and B. T. Murray. "Tolerating Transient Faults in Statically Scheduled Safety-Critical Embedded Systems". In *Proc. of the 18th IEEE Symposium on Reliable Distributed Systems (SRDS)*, pages 212–221, 1999.
9. R. M. Kieckhafer, C. J. Walter, A. M. Finn, and P. M. Thambidurai. "The MAFT Architecture for Distributed Fault Tolerance". *IEEE Transactions on Computers*, 37(4):398–405, April 1988.

10. H. Kopetz. *Real-Time Systems Design for Distributed Embedded Applications.* Kluwer Academic Publishers, 1997.
11. C. M. Krishna and K. G. Shin. "On Scheduling Tasks with a Quick Recovery from Failure". *IEEE Transactions on Computers*, 35(5):448–455, 1986.
12. F. Liberato, R. G. Melhem, and D. Mossé. "Tolerance to Multiple Transient Faults for Aperiodic Tasks in Hard Real-Time Systems". *IEEE Transactions on Computers*, 49(9):906–914, 2000.
13. L. Liestman and R. H. Campbell. "A Fault-Tolerant Scheduling Problem". *IEEE Transaction on Software Engineering*, 12(11):1089–1095, 1986.
14. G. M. A. Lima. *"Fault Tolerance in Fixed-Priority Hard Real-Time Distributed Systems"*. PhD thesis, Deptartment of Computer Science, University of York, 2003.
15. G. M. A. Lima and A. Burns. "An Optimal Fixed-Priority Assignment Algorithm for Supporting Fault Tolerant Hard Real-Time Systems". *IEEE Transaction on Computers*, 52(10):1332–1346, 2003.
16. S. Poledna. *Fault-Tolerant Real-Time Systems: The Problem of Replica Determinism.* Kluwer Academic Publishers, 1996.
17. S. Poledna, A. Burns, A. J. Wellings, and P. Barrett. "Replica Determinism and Flexible Scheduling in Hard Real-time Dependable Systems". *IEEE Transsactions on Computers*, 49(2):100–111, 2000.
18. S. Punnekkat. *"Schedulability Analysis for Fault Tolerant Real-Time Systems"*. PhD thesis, Department of Computer Science, University of York, 1997.
19. S. Ramos-Thuel and J. K. Strosnider. "The Transient Server Approach to Scheduling Time-Critical Recovery Operations". In *Proc. of the 12th Real-Time Systems Symposium (RTSS)*, pages 286–295. IEEE Computer Society Press, 1991.
20. F. Schneider. "Replication Management Using the State-Machine Approach". In Sape Mullender, editor, *Distributed Systems*, chapter 7. Addison-Wesley, 2nd edition, 1993.
21. P. Veríssimo and L. Rodrigues. *Distributed Systems for Systems Architects.* Kluwer Academic Publishers, 2001.

On the Monitoring Period for Fault-Tolerant Sensor Networks*

Filipe Araújo and Luís Rodrigues

Universidade de Lisboa, Departamento de Informática, Faculdade de Ciências,
Campo Grande, Edifício C6, 1749-016 Lisboa, Portugal
{filipius, ler}@di.fc.ul.pt

Abstract. Connectivity of a sensor network depends critically on tolerance to node failures. Nodes may fail due to several reasons, including energy exhaustion, material fatigue, environmental hazards or deliberate attacks. Although most routing algorithms for sensor networks have the ability to circumvent zones where nodes have crashed, if too many nodes fail the network may become disconnected.

A sensible strategy for increasing the dependability of a sensor network consists in deploying more nodes than strictly necessary, to replace crashed nodes. Spare nodes that are not fundamental for routing or sensing may go to sleep. To ensure proper operation of the sensor network, sleeping nodes should monitor active nodes frequently. If crashed nodes are not replaced, messages follow sub-optimal routes (which are energy inefficient) and, furthermore, the network may eventually become partitioned due to the effect of accumulated crashes. On the other hand, to save the energy, nodes should remain sleeping as much as possible. In fact, if the energy consumed with the monitoring process is too high, spare nodes may exhaust their batteries (and the batteries of active nodes) before they are needed.

This paper studies the optimal monitoring period in fault-tolerant sensor networks to ensure that: *i)* the network remains connected (i.e., crashed nodes are detected and substituted fast enough to avoid the network partition) and, *ii)* the lifetime of the network is maximized (i.e., inactive nodes save as much battery as possible).

1   Introduction

Sensors have long since been used for monitoring processes where humans are either endangered by hazardous environments, too costly to be an option, or simply not able to effectively perform the sensing task. Recent progresses in miniaturization and networking technologies are empowering the use of sensors in self-organizing wireless networks, where nodes cooperate to more effectively achieve some goal. Wireless sensor networks have a wide range of applications, such as military, commercial, industrial, home or health.

* This work was partially supported by the LaSIGE and by the FCT project P-SON POSC/EIA/60941/2004 via POSI and FEDER funds.

C.A. Maziero et al. (Eds.): LADC 2005, LNCS 3747, pp. 174–190, 2005.

In this paper we study techniques to increase the dependability of sensor networks. Nodes that crash can reduce the accuracy or completeness of the information being collected. Additionally, if too many nodes fail, the network may become disconnected. Therefore, we are particularly concerned with techniques that extend the lifetime of the network by postponing disconnection. A sensible strategy for increasing the dependability of a sensor network consists in deploying more nodes than strictly necessary. In this way, nodes collectively decide which ones remain active and which ones may go to sleep. To ensure proper operation of the sensor network, sleeping nodes should monitor active nodes frequently. Crashed nodes may cause sub-optimal routing (which wastes energy) as well as a network partition. On the other hand, to save the energy, nodes should remain sleeping as much as possible. If the energy consumed with the monitoring process is too high, spare nodes may exhaust their batteries (and the batteries of active nodes) before they are needed.

In this context, we would like to select a value for the monitoring period that maximizes the system availability. This task can be prohibitively complex due to the multiple combinations of factors that affect the system lifetime such as the initial energy available to nodes, power consumption, network topology, etc. The paper addresses this complexity by making the following contributions: in first place, it proposes a methodology of analysis that simplifies the reasoning about the network behavior and, in second place, it proposes two new metrics that capture the importance of the *relative* values of different system parameters. The first metric, called "Failure Weight Factor", $\mathcal{F}$, relates the Mean Time Between Failures, $MTBF$, with the maximum lifetime of the network, in ideal monitoring conditions. The second metric, called "Power On-off Consumption Factor", $\mathcal{P}$, relates the energy spent powering nodes on and off with the energy spent by other sources of energy consumption. Using simulations, we show that these two metrics are useful to reason about the impact of faults in the network lifetime.

The rest of the paper is structured has follows. Section 2 overviews related work. Section 3 presents our reference cell-based algorithm for energy conservation. Section 4 describes our metrics and the analysis methodology. The simulation results are presented and discussed in Section 5. Finally, Section 6 concludes the paper.

2   Background

The benefits gained from having more nodes than necessary have to be balanced against the (energy) costs of managing the nodes. In this section, we overview related work that helps to answer the following questions: How can the lifetime of a sensor network be precisely defined? How is energy consumed in a sensor network? Which are the best techniques to tolerate node failures? Should some redundant nodes be kept idle or, on the contrary, should all redundant nodes be kept sleeping most of the time? How to replace nodes whose energy has been exhausted? Which routing algorithm should be adopted? Previous work on the above topics helps us to define our strategy to build a fault-tolerant sensor network.

Network Lifetime. In the literature, there are several definitions of network lifetime [25,30,7], like time to first node that dies. In this paper, we adopt the definition from [3], which considers that network life ends when the first partition occurs. For the scenarios considered in this paper, this metric offers a good measure of the availability of the network, because partitions typically occur little after half of the space where the sensor network lies becomes empty of nodes[1].

Energy Consumption. In a sensor network, tasks that typically consume more energy are: sending and receiving messages, listening to the channel when idle, and processing. In this paper we do not consider sensing energy, because it depends mainly on the sensing task.

Several papers report that nodes consume a significant amount of energy in idle mode [23,12,8]. According to [12], the ratio of power needed in receive (transmit) mode against idle mode can be as low as 1.15 (1.56)[2]. This order of magnitude for idle power consumption paves the way to selectively powering down nodes, to conserve energy, because nodes consume only a small amount of energy while sleeping. One aspect that is often overlooked in literature is the cost of powering on and off a node. We believe that any algorithm that selectively powers down nodes to save energy must address this cost. In fact, there are two issues to consider: the time it takes to wake up and the large spike in energy consumption, due to the wakeup action alone plus a traffic announcement. The exact figures for both of these depend on the communication card and controlling software.

Fault-Tolerant Wireless Networks. Resilience to node failures and energy efficiency must be addressed simultaneously, because an energy-efficient routing algorithm should be fault-tolerant and fault-tolerance can not come at a high energy cost. For this reason, several authors focused on algorithms that are both fault-tolerant *and* energy efficient (e.g, [10] and [11]). Several authors propose heuristics to ensure k-vertex connectivity [13,5,17]. Unfortunately, this construction requires nodes to be active when they are not strictly required. The amount of energy consumed this way results in an effective loss of network lifetime. Therefore, most approaches to extend the network lifetime try to power down redundant nodes.

Powering Down Nodes. There are many protocols that explore the idea of powering down redundant nodes, both at network and MAC layers (e.g. [22]). Attacking the problem at the network layer typically enables longer sleeping periods, because decisions are more informed. Instead of powering down a node for a single message or for some predefined number of time slots, knowledge of the routing algorithm can be used to selectively put some (almost) redundant nodes to sleep. For instance, this is the case of Span [8] that allows some nodes to sleep if node density is high enough. In [26], authors address the integration of the connectivity with the coverage problem. Other authors propose to selectively power down nodes in cluster-based routing schemes [30,29,28]. In cluster-based algorithms, a good policy to select the cluster-heads is the available energy (e.g., [28,29]), instead of other criteria,

---

[1] More precisely, most partitions occur after half of the network cells become inactive. A precise definition of the concept of cell can be found ahead in this text.

[2] In a Lucent IEEE 802.11 WaveLan PC Card.

like node id (e.g., [27]), or node degree (e.g., [9]). Due to their very well-structured organization and predictable behavior, division of the space into cells constitutes the ideal scenario to analyze the impact that the monitoring period has on the lifetime. In fact, we will show that this division allows us to evaluate precisely the effect of each input variable on the lifetime and, consequently, on the ideal monitoring period setting. For this reason, in this paper, we will adopt a modified version of Geographical Adaptive Fidelity, GAF [30], which we present in Section 3.

Routing Algorithm. Several proposals for energy-aware routing strategies can be found in the literature [21,6,25]. While some of these strategies aim to prolong as much as possible the lifetime of the first node to die [6,25], others try to avoid the exhaustion of the entire network [21]. To reduce power needed to transmit, nodes might adjust their transmission range. Using this technique, two papers [25] and [2] showed that the best strategy to deliver a message over a total distance D is to use equally spaced hops. Although in practice networks do not have nodes ideally located to relay a message, this result allows to derive upper bounds on network lifetime [2] and to build power-aware routing algorithms [25]. In [7,25] authors simultaneously try to minimize power consumption as a whole and avoid exhaustion of nodes short of energy. Often, avoiding individual node depletion is not an issue in a sensor network, where fairness is less important than maintaining the network functioning.

The use of positional information is also important to conserve energy. As pointed out in [24,14], positional routing algorithms make a more efficient use of resources than other routing algorithms like AODV [18], DSDV [19] or DSR [15] in large networks, because they use much fewer control messages. Additionally, positional information for the routing algorithm, in a scenario where a cell-based conserving energy algorithm is in use, comes for free, because a GPS receiver or an equivalent mechanism already exists. These facts motivated us to use, for the purpose of this study, a position-based routing algorithm. By avoiding algorithms that require configuration of several parameters, we also avoid the risks of having our results biased by inappropriate settings.

Therefore, we selected the Greedy Perimeter Stateless Routing (GPSR) algorithm [4,16], because it is localized, and efficient. Furthermore, since GPSR has a very simple configuration and very few dedicated control messages, its operation has very little interference in our results. When possible, GPSR uses the greedy strategy of forwarding messages to the neighbor closest to destination. When it finds a local minimum, GPSR switches to perimeter mode and routes around faces. As soon as it finds a node closest to destination than the previous local minimum, GPSR goes back to greedy mode.

3   An Approach to Build Fault-Tolerant Sensor Networks

To build a fault-tolerant sensor network we include more nodes than strictly required. This allows replacement of failed nodes. To save battery, nodes collectively decide which ones are not fundamental for routing or sensing. These nodes should be sleeping most of the time, and only wake up with the minimum frequency required to replace failed nodes before the network disconnects. There are several issues that have to be defined in order to implement this strategy. In first place, nodes have to agree on some strategy to define which nodes should sleep, and which nodes must remain in idle state to maintain the network connectedness. In second place, one needs to define a strategy to perform the monitoring of idle nodes. Finally, one needs to define how often the monitoring procedure should be performed. This paper tackles the latter two issues, with particular emphasis on the importance of the monitoring period. As motivated in the previous section, we base our architecture in a GAF [30]-like cell based network running GPSR [16].

3.1   Node Monitoring in Geographical Adaptive Fidelity

Geographical Adaptive Fidelity (GAF) [30] is a cell-based energy-conserving algorithm. GAF aims to maintain all but one node sleeping in each cell. It assumes that nodes are aware of their location (for instance, using GPS receivers) and uses this information to divide the two-dimensional space into a grid. The two farthest points in any two adjacent cells must be within communication range, as depicted in Figure 1a. This bounds the cell side, r, to $r \leq R/\sqrt{5}$, where R is the communication range of the nodes. In scenarios where it is worthwhile using GAF, because more than one node exists per cell, resulting graph is very likely to be connected.

In GAF, nodes can be in one of three states: *active*, *discovery* or *sleeping*. Changes from one state to another are controlled by *discovery* messages and by timers. A node uses *discovery* messages to inform other nodes of its presence and of its application-dependent rank. In [30], authors propose as a ranking criterion, first, the state of the node (active > discovery) and then the expected lifetime, *enat* (higher ranks correspond to longer expected lifetimes). Hence, *discovery* messages consist of the following tuple: {node id, grid id, estimated node active time (*enat*), node state}. As depicted in Figure 1b, timers can change state of a node from sleeping to discovery (after T_s), from discovery to active (after T_d) and from active to discovery (after T_a). Nodes send discovery messages in any of the following situa-

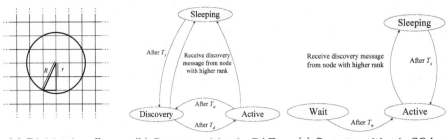

(a) Division in cells   (b) State transition in GAF   (c) State transition in SQA

Fig. 1. GAF and SQA algorithms

tions: i) when they enter the discovery state, ii) when they enter active state after timeout T_d takes them from discovery to active; iii) periodically, after each period of T_d seconds in active state; iv) in active state when they receive a *discovery* message from a node with lower rank. Whenever a node in discovery or active states receives a *discovery* message from a node with higher rank it immediately resets its ongoing timers, sets up a timer to wake up and changes to sleeping state.

If nodes are put to sleep for too long, it may happen that the node occupying the cell either exhausts its battery or abandons the cell (if it is mobile) leaving it unattended. On the other hand, if sleeping nodes wake up too early, they will consume everybody's resources without further improving routing fidelity, thus defeating the goal of maximizing network lifetime. To achieve a good trade-off, GAF dynamically sets the sleeping period of a node, T_s, to depend on the estimated lifetime of the cell leader. In GAF, T_s is set to a fraction (50%) of the estimated lifetime of the leader. Hence after the T_a timer of the leader expires, it switches from active to discovery state, thus having an opportunity to be replaced in the cell. This is important for load balancing purposes (see [30] for further details).

3.2 Sleep-Query-Active Algorithm

Unlike [30], in this paper we consider some additional characteristics that make a more realistic scenario: i) nodes can fail and ii) waking up and putting nodes to sleep has fixed non-negligible cost. Furthermore, since we only consider sensor networks of fixed nodes, load balancing is not an issue. These differences motivated us to develop a variation of the GAF algorithm, which we call Sleep-Query-Active Algorithm (SQA), specifically suited to our setting. The states of SQA are depicted in Figure 1c. SQA is a pretty simple algorithm where nodes can only be in one of two states: either sleeping or active. The purpose of the *wait* state is only to desynchronize nodes that start at the same time. In our experiments, T_w was randomly set between 0 and 1 with uniform probability.

SQA nodes send discovery messages in the following situation: i) when they enter active state, ii) periodically when they are in the active state (to overcome the loss of messages) and iii) in active state when they receive a *discovery* message from a node with lower rank. Differences to GAF in the exchange of *discovery* messages, mainly reflect the way the rank is determined. In SQA the rank of the node is determined by the *enat* alone. Despite not providing any additional protection against node failures, nodes with larger supplies of energy will give an additional degree of protection against unexpected energy consumption caused by some peak of traffic. Perhaps the most important difference between GAF and SQA is that in SQA the sleeping timeout, T_s, which we deem as the *monitoring period*, is randomly chosen from an interval that is fixed beforehand. When we say that $T_s = c$, we really mean that T_s is selected from the interval $[0.5 \times c, 1.5 \times c]$. Then, each time a node goes to sleep, it picks the value for T_s from that interval with uniform probability. Our experimental evaluation shows that this choice is appropriate, because more often that not, the sensor networks will tend to behave in a very predictable way and using an optimal fixed value for T_s will yield longer lifetimes than the dynamic approach of GAF. The reader should notice that tuning SQA resumes to determining T_s. Selecting the most appropriate T_s is a challenging task that we address in the next sections. In fact, as we show in Section 5, for an appropriate choice of the monitoring period, SQA can successfully replace GAF in sensor networks.

4   Proposed Metrics and Analysis Methodology

When using SQA, we would like to determine the monitoring period T_s that maximizes network lifetime. Unfortunately, following a theoretical approach to determine T_s is a task of great difficulty. An example of such an attempt can be found in [3], where a theoretical bound for the network lifetime in a scenario where dead nodes are replaced at once without spending energy (we will call this the "ideal scenario" or "ideal network") is derived. However, that work does not account for all the parameters we consider in this paper (e.g. faults) and, as noted in [3], it cannot be easily extended to capture practical scenarios. Hence, in this paper we have opted to use simulations to evaluate the effect of different parameters on the T_s. Unfortunately, without a correct methodology, the process of determining the effect of T_s on a network using simulations is also a daunting task. In fact, there are many factors that can influence network lifetime and consequently, T_s, including initial energy of nodes, idle energy consumption, transmission power, consumption power, sleep energy consumption, not to mention power on consumption and faults. Furthermore, these factors can be combined in multiple ways and often can not be completely isolated in order to analyze their impact on network lifetime. Finally, but not the least, a single ns-2 [1] simulation of a given configuration (i.e., for a single monitoring period), even when in executed on a Pentium IV 2.8 GHz with 2Gb of RAM, takes more than 100 seconds to complete.

To handle this complexity, the paper makes two contributions. In first place we propose a new set of metrics to reason about the influence of faults in the network lifetime. An interesting feature of these metrics is that they capture the *relative* weight of different factors, and highlight that networks with different absolute values of some parameters may exhibit a comparable behaviour. In second place, we propose a methodology of analysis that allows to reason about the impact of these metrics before assessing the impact of network topology in the final system availability. We will address these two contributions in the following paragraphs.

4.1   The $\mathcal{P}$ and $\mathcal{F}$ Metrics

Our metrics are motivated by the insight that, in the context of assessing the network availability, time intervals – in particular the monitoring period – should be analyzed in a relative sense: a monitoring period of 1 second has a different impact on a network whose lifetime is just 10 seconds than on a network whose lifetime is 1000 seconds. In a similar manner, the magnitude of values like power needed to transmit or to receive should also be measured in a relative way.

To reason in a generic manner about the fault-tolerance and power-on consumption of sensor networks, we start by defining the notion of *ideal lifetime*, LT_I. LT_I is the network lifetime in a scenario where i) there are no faults, ii) switching nodes on and off has no cost and *iii*) nodes in the cells are omnisciently replaced at once (if replacement is available). LT_I is determined by simulation and measures the available initial energy versus average consumption of the network. Using LT_I we propose the following metrics to assess the network behavior:

- The *power on-off consumption factor*, $\mathcal{P}$, measures the impact of the energy spent powering nodes on and off. We define it as the ratio between the energy needed for one power on-off operation versus remaining energy spent in 1 time unit. This is determined as $\mathcal{P} = POE/(TE_0/LT_I)$, where POE is the power on-off energy and TE_0 is the total energy available in the beginning of the network life (if we assume that all N nodes have the same energy, E_0, in the beginning, $TE_0 = N \times E_0$). This makes $\mathcal{P}$ a function of all remaining energies of the system *but not* of node failure rate.
- The *failure weight factor*, $\mathcal{F}$, measures the impact of faults in the network. We define it as the lifetime of the ideal network, LT_I, relative to $MTBF$, i.e., $\mathcal{F} = LT_I/MTBF$. This makes $\mathcal{F}$ a function of all energies *except* power on-off energy. Large $\mathcal{F}$ means many node failures (possibly due to a long network lifetime), while large $\mathcal{P}$ means a lot of energy needed to power a node on and off (at least compared with remaining energies, like idle and traffic energies).

4.2   From Cell Level to Network Level Simulation

We propose and use the following methodology to evaluate the lifetime of the wireless network. Instead of always running simulation on a complete network, we first perform a careful study of the behavior of each network cell. Then, by estimating how many cells are required to maintain the connectivity of a given topology, we extrapolate the impact of the parameters in the entire network. We illustrate this methodology in Figure 2. The approach has both conceptual and practical advantages. From the conceptual point of view, it allows to separate the analysis of the influence of topology from other factors. From the practical point of view, cell level simulations *i*) allow to isolate factors that influence network lifetime and *ii*) run much faster. Therefore, cell simulation allows a much richer analysis of different combinations of factors in practical time. We validate our methodology by comparing the results obtained using this method with the results obtained by simulating the entire network. An additional advantage of the cell simulations is that its results can be used to assess other system properties. For instance, although outside

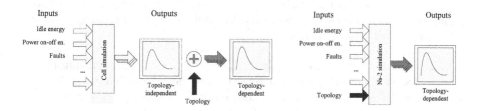

Fig. 2. Cell Based Methodology *vs* Network Simulation

the scope of this paper (where we focus on network lifetime) the analysis of cell simulations could be easily extended to study the problem of assessing the coverage of the sensor network in presence of faults.

5   Experimental Results

In this section we present our simulation results. We start by describing the settings used to perform cell level simulations and network level simulations. We then validate our methodology by comparing results derived from it (based on cell level simulations) with results obtained by directly simulating the entire network. Later, we show the relevance of the $\mathcal{P}$ and $\mathcal{F}$ metrics and their impact on the network lifetime. Finally, we illustrate the importance of appropriately selecting the correct monitoring period.

5.1   Simulation Settings

In our experiments we have used three different types of nodes: a node equipped with a Lucent IEEE 802.11 2 Mbps WaveLAN PC Card, a Rockwell's WINS node and a MEDUSA-II node. Table 1 resumes the consumption of the three different nodes in the situations considered in our simulations. Figures for the first node were taken from [12], while values for the other two types of nodes were inferred from [20].

We assume that failures of nodes follow an exponential distribution. However, for simulation purposes, we have modeled this as a geometric distribution. After constant time intervals P, all nodes may fail with a given random probability p (we set $P = 0.5$ seconds in our simulator). Hence parameter r of the exponential distribution is $r \cong -\frac{1}{P}\ln(1 - p)$, while $MTBF = 1/r$.

To plot a graphic that represents lifetime relative to LT_I against the monitoring period relative to LT_I (e.g., Figure 4), we select a number of monitoring periods, T_s, not exceeding the ideal lifetime. Then, we fix all the parameters, like power on-off consumption, idle power, initial energy, etc. and we experimentally analyze the lifetime achieved for each T_s. We used a square size of 800 × 800 meters with 256 nodes, which we divided into 8 × 8 squares (giving an average of 4 nodes per cell). Communication range was 250 meters. The main difference between the cell and the ns-2 experiments is the way in which lifetime is determined. In ns-2 we run a simulation of the entire network to determine this value, while in the cell simulations we use a method that we describe next. We have performed additional simulations that show that these results also apply when other topologies are used (this aspect is discussed in Section 5.5).

Cell Level Simulation Settings. To determine the lifetime for a given monitoring period, we fix this monitoring period and use time as the independent variable. Then, as time goes by we assume a constant consumption of energy and observe whether the cell is awake or sleeping (it is awake if there is any node awake, otherwise it is sleeping). We used an average of 100 of these trials to approximate a continuous random variable, function of time t, that represents the probability that the cell

Table 1. Consumption of energy for the nodes tested

Node	Rx (W)	Tx (W)	Idle (W)	Sleep (W)	Initial Energy (J)
IEEE 802.11	0.974072	1.3410736	0.843	0.066303	15
MEDUSA-II	0.01248	0.01565	0.01234	0.00002	1
Rockwell's WINS	0.7516	1.0805	0.7275	0.064	20

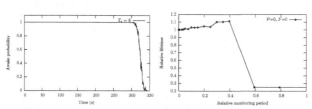

(a) Probability of a cell being awake

(b) Relative network lifetime

Fig. 3. Derivation of network lifetime in cell simulation

is awake. An example of a random variable like this is depicted in Figure 3a, for a specific value of T_s. To infer network behavior from this, we need to know the topology of the network. If disconnection occurs when an average number of D out of N cells are sleeping, we use a rough approximation and assume that when the awake probability of a cell drops below $(N - D)/N$, the network gets disconnected. Taking our grid for example, we used a simple simulation to derive the probability density function of the number of sleeping cells that cause network disconnection. This looks like a Gaussian curve centered at 40 and truncated at the 64 cells. Therefore, in such a topology, the threshold $(64 - 40)/64 = 0.375$ corresponds to a point where, more often that not, network will be disconnected [3]. Figure 3b shows the relative lifetime graph as a function of the monitoring period for these settings.

Lifetime and monitoring periods represented in this plot are relative to the ideal lifetime LT_I, to abstract away the absolute magnitudes that govern the network behavior. Note that an entire data series needed to create a graphic like the one represented in Figure 3a produces a single point in Figure 3b. In this case, this point should occur around $t = 327$ seconds (where the line $y = 0.375$ intersects the probability curve). In the cell simulations LT_I is estimated as the number of nodes of the cell × the time it takes to consume all the energy of a node [4]. For the settings of these figures, this is around 324. Since $T_s = 8$ and $LT = 327$, this gives a relative monitoring period of $8/324 \approx 0.025$ and a relative lifetime of $327/324 \approx 1.009$. It is not really counterintuitive to have a lifetime greater than the ideal, due to the large idle power. In fact, this makes it advantageous to let some cells sleeping from time to time, to prolong their lives. On the contrary the ideal lifetime assumes that all the cells should be constantly awoken, which is not always the best strategy.

---

[3] In this case, disconnection occurs when a significant proportion of the network is, in fact, unusable. We also observed this for other grid configurations.

[4] The reader should keep in mind that this only refers to the cell simulations.

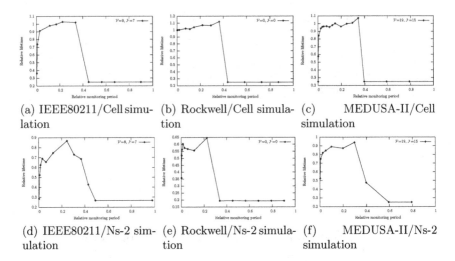

(a) IEEE80211/Cell simulation (b) Rockwell/Cell simulation (c)   MEDUSA-II/Cell simulation

(d) IEEE80211/Ns-2 simulation (e) Rockwell/Ns-2 simulation (f)   MEDUSA-II/Ns-2 simulation

Fig. 4. Lifetime estimated using cell level and network level simulations

Network Level Simulation Settings. We used the ns-2 simulator [1], version 2.27, to perform the network level simulations presented in this paper. This required us to implement the SQA algorithm as well as port the GPRS routing algorithm to the same version of ns-2. We used a simulation environment similar to the one described in [30]. Nodes were divided in traffic and transit nodes. Traffic nodes serve as sources and sinks of traffic, while transit nodes are only used as intermediate hops for that traffic. Only transit nodes run the SQA/GPSR protocol. Traffic was generated by constant bit rate (CBR) traffic sources. In all our experiments we fixed the number of traffic nodes to 10. To prevent traffic nodes to stop generating traffic, their supply of energy was infinite.

5.2   Validation of the Methodology

To validate our methodology, we compare the results obtained from the application of the methodology, with the results obtained from complete network level simulations, using ns-2. Samples of several simulation we have performed for three different concrete node characteristics are depicted in Figure 4. Although the shape of the lines is slightly different, the peak in the relative lifetime plots is comparable, despite huge differences in power figures of nodes. This is very important, because in this peak lies the answer to the main question of this paper: what is the optimal selection of T_s? The fact that its width is similar in both types of simulations, allows us to use the simpler cell simulations to reason about the impact of the $\mathcal{P}$ and $\mathcal{F}$ metrics.

5.3   Relevance of $\mathcal{P}$ and $\mathcal{F}$ Metrics

Impact of the Power Parameters on the Lifetime We observed that the impact of the power parameters, like idle, transmission or reception, can be hidden by plotting curves relative to the ideal lifetime. This was a surprising result of our simulations. Experiments made both with cell level simulations and in ns-2 confirmed this observation. Figure 4 allows to confirm this, because the three types of nodes have similar curves despite the differences in their power ratings (note for instance that MEDUSA-II consumptions are orders of magnitude away from the other types of nodes). Hence, the effect of the absolute values of the power consumptions are almost entirely ruled out, by using the simple technique of plotting lifetime curves relative to the ideal lifetime. This considerably simplifies the analysis of the metrics $\mathcal{P}$ and $\mathcal{F}$ to be done ahead.

The parameters that have larger impact on relative network lifetime curves are the power on consumption (assessed by $\mathcal{P}$) and the faults (assessed by $\mathcal{F}$). Impact of the node density is discussed in Section 5.4.

Impact of the Metrics $\mathcal{P}$ and $\mathcal{F}$ on the Lifetime. We now use cell level simulations to discuss the impact of faults (represented by $\mathcal{F}$) on the network lifetime considering a non-negligible replacement cost (represented by $\mathcal{P}$). For most values of $\mathcal{P}$ and $\mathcal{F}$, the stability of the lifetime peak still holds. Since several combination of input parameters are captured by the two metrics, a precise determination of these metrics should be enough to qualitatively determine the behavior of the network. Figure 5 shows extreme as well as typical values for $\mathcal{P}$ and $\mathcal{F}$. We can see that results confirm the initial intuition: large values of $\mathcal{F}$ tend to require smaller monitoring periods (thus shrinking the curve at the right and making the peak start slightly earlier). On the other hand, larger values of $\mathcal{P}$ will penalize small monitoring periods (thus shrinking the curve at the left). Hence, as these two metrics grow, the curve tends to become thinner. Moreover, the growth of these metrics also makes the curve shorter as they impact network lifetime. To conserve space we only depict results for the IEEE 802.11 adapter. However, results for the other types of nodes show similar behaviors. Together with other simulations that we have done, this shows that very different operational conditions have similar behaviors, as long as the metrics $\mathcal{P}$ and $\mathcal{F}$ are similar (this effect also occurs in Figure 4).

Table 2, which summarizes the results obtained, offers a qualitative analysis of this issue. Outside the parenthesis we describe the system parameter that dominates network lifetime (other energies refers to idle and traffic energies), while inside we describe the shape of the peak that exists in the monitoring period (earlier, normal or later, respectively means that peak starts closer, in the normal place or farther away from the origin). Given the values of Table 1 and the huge idle mode power, we expect current technology to operate in the first line of the table ("Small $\mathcal{P}$"). If with technological improvements idle energy decreases, $\mathcal{P}$ will depend mainly on data traffic generated on the network. In this case, the network will operate in a zone captured by the bottom line of the table ("Large $\mathcal{P}$"), whenever average traffic becomes low. In such scenarios, the appropriate choice of T_s will make an even more significant impact on the network lifetime.

In our simulations, including results depicted in Figures 4 and 5, longest lifetimes are almost always achieved when monitoring period is in the range of 10 to 20% of the ideal lifetime, for most values of $\mathcal{P}$ and $\mathcal{F}$. This stability has to do with the fact that a perfect monitoring algorithm should ensure that network has as few

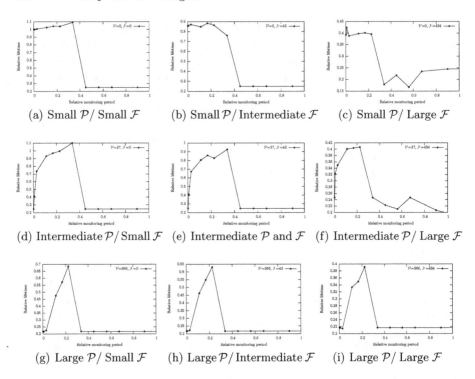

Fig. 5. Impact of $\mathcal{P}$ and $\mathcal{F}$

Table 2. Dominating parameter (and peak shape) for variations of $\mathcal{F}$ and $\mathcal{P}$

	Small $\mathcal{F}$	Intermediate $\mathcal{F}$	Large $\mathcal{F}$
Small $\mathcal{P}$	Other en. (earlier)	Other en. & Failures (earlier)	Failures (earlier)
Intermediate $\mathcal{P}$	All en. (normal)	None (normal)	Failures (slightly earlier)
Large $\mathcal{P}$	On-off (later)	On-off (later)	Depends rel. magnitude (later)

active nodes as possible (fewer than the number of cells, in practice), but preserving the minimum required to prevent disconnection from occurring. Hence, substitution of nodes depends on the rate nodes die, which on its turn will determine lifetime. This explains why better strategies for (potentially) longer lifetimes, should use longer monitoring periods. Nevertheless, if this period goes over some threshold (30 to 50%), the relative lifetime sharply decreases, because nodes that die are not replaced and many cells become empty. This reveals a thin line between optimal and disastrous configuration.

5.4   Impact of Node Density on the Lifetime

One aspect of our results that is difficult to understand with the ns-2 simulations, but evident in cell simulations is the impact of node density. Cell experiments (that we omit to conserve space), have shown that the peak of the lifetime curve shrinks when the number of nodes per cell increases. This is consistent with results obtained in ns-2 (IEEE 802.11) and depicted in Figure 6a, where this effect is quite subtle. In this experiment we fixed all parameters and varied the number of nodes from 64 to 512 (density $d = 1$ represents 256 nodes). The gain in lifetime (relative to the lifetime of density 1) is depicted in Figure 6b for different network densities. We have studied two scenarios of independent interest: ideal replacement policy with and without node failures. The approximately linear growth of lifetime when there are no failures is consistent with [3]. However, when we consider failures of nodes, as absolute lifetime increases, failures become more important ($\mathcal{F}$ grows). This makes lifetime (relatively) shorter as density increases.

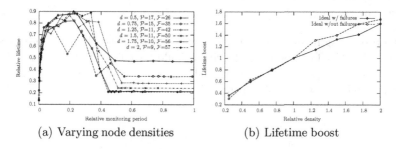

(a) Varying node densities   (b) Lifetime boost

Fig. 6. Relative lifetime and lifetime boost with varying node densities

5.5   Impact of Topology on the Lifetime

Other experiments that we have made with ns-2 for the topology settings originally described in [30] did not show significant changes to the results presented here. This scenario is of particular interest, because nodes are scattered in a rectangle of 1500×300 meters with 100×100 meters for each cell, which gives only 3 cells in one of the directions. Cell simulations with thresholds different from 0.375 also have also produced similar results. Nevertheless, we believe that it is still an open problem to know if there are configurations that considerably impact the lifetime of the network and how can that impact be predicted.

5.6   Practical Relevance

We finally show in Figure 7 the benefit from adequately selecting the monitoring period T_s. We illustrate this by using several different replacement policies in scenarios with increasing node failure rates, simulated for 256 nodes in ns-2 (IEEE 802.11 adapter). $\mathcal{F} = 0$ means that there are no failures, i.e., $MTBF = \infty$. First, we determine an upper bound for the lifetime using an ideal scenario with node failures ("Ideal w/ failures"). Next, we use a worst-case setting where T_s is so long that no actual substitution ever occurs ("Pessimal"). The third intermediate scenario consists of keeping all nodes awake. In this case, no idle energy is conserved ("All active"). The y-axis of the graphic is normalized to the ideal lifetime, LT_I (which does not vary along the x-axis, as it does not have node failures).

Then we plot two additional curves in the graphic: lifetime obtained by the GAF algorithm and lifetime obtained by SQA. For SQA we select the monitoring period using the results from the analysis presented in Subsection 5.3: we selected smaller monitoring periods for larger values of $\mathcal{F}$, starting at 20% of LT_I, for small values of $\mathcal{F}$ and decreasing for 15%, 10% and finally 5% as $\mathcal{F}$ grew larger. From the figure we can reach the following conclusions:

- Not adjusting the monitoring period (for instance, using the pessimal or the all active approaches) offers a network lifetime that is much worse than the ideal.
- Using the analysis presented in this paper, SQA can be tuned to achieve a lifetime that is frequently between 80 and 90% of the ideal.
- SQA offers, for most values of $\mathcal{F}$, a much longer network lifetime than GAF, that can be as high as 25%.

As a promising future research topic, we envision to combine the advantages of SQA and GAF. The resulting algorithm could have the ability to dynamically set the monitoring period, according to the importance of faults existing on the network or to the power on-off consumption.

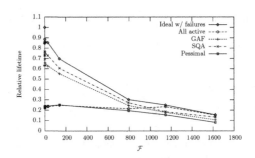

Fig. 7. Lifetime for Different Replacement Methods

6   Conclusions

In this paper we studied the dependability of sensor networks, considering energy constraints and fault-tolerance requirements. We aimed at determining the ideal monitoring period for cell-based energy conserving techniques, to maximize network lifetime, here defined as time to the first network partition. To simplify this task, this paper made two contributions: a methodology of analysis, which consisted of inferring network behavior from inspection of individual cells; and two metrics, $\mathcal{P}$ and $\mathcal{F}$ that are able to capture the operational conditions of the sensor network.

Experimental results demonstrated the appropriateness of using these metrics to assess network behavior, by showing that, often, $\mathcal{P}$ and $\mathcal{F}$ strongly determine network operation. Furthermore, results have shown that it is possible to achieve a lifetime close to the ideal by selecting the monitoring period adequately and according to $\mathcal{P}$ and $\mathcal{F}$. More precisely, we have shown that network lifetime can be within 80 and 90% of that provided by an (non-implementable) ideal replacement policy, even for very large failure rates.

References

1. *The* ns *Manual.* http://www.isi.edu/nsnam/ns/ns-documentation.
2. M. Bhardwaj, A. Chandrakasan, and T. Garnett. Upper bounds on the lifetime of sensor networks. In *IEEE International Conference on Communications*, pages 785–790, 2001.
3. D. Blough and P. Santi. Investigating upper bounds on network lifetime extension for cell-based energy conservation techniques in stationary ad hoc networks. In *ACM Mobicom*, 2002.
4. P. Bose, P. Morin, I. Stojmenović, and J. Urrutia. Routing with guaranteed delivery in *ad hoc* wireless networks. In *International Workshop on Discrete Algorithms and Methods for Mobile Computing and Communications (DIALM)*, pages 48–55, 1999.
5. G. Calinescu, I. Mandoiu, and A. Zelikovsky. Symmetric connectivity with minimum power consumption in radio networks. In *17th IFIP World Computer Congress*, pages 119–130, 2002.
6. J. Chang and L. Tassiulas. Routing for maximum system lifetime in wireless ad-hoc networks. In *37-th Annual Allerton Conference on Communication, Control, and Computing*, Monticello, IL, September 1999.
7. J.-H. Chang and L. Tassiulas. Energy conserving routing in wireless ad-hoc networks. In *INFOCOM (1)*, pages 22–31, 2000.
8. B. Chen, K. Jamieson, H. Balakrishnan, and R. Morris. Span: An energy-efficient coordination algorithm for topology maintenance in ad hoc wireless networks. *Wireless Networks*, 8(5):481–494, 2002.
9. G. Chen and I. Stojmenovic. Clustering and routing in wireless ad hoc networks. Technical Report TR-99-05, Department of Computer Science, SITE, University of Ottawa, Ottawa, Ontario K1N 6N5, Canada, June 1999.
10. S. Chessa and P. Santi. Crash faults identification in wireless sensor networks. *Computer Communications*, 45(2):126–143, November 2002.
11. A. Datta. Fault-tolerant and energy-efficient permutation routing protocol for wireless networks. In *International Parallel and Distributed Processing Symposium (IPDPS'03)*, 2003.
12. L. M. Feeney and M. Nilsson. Investigating the energy consumption of a wireless network interface in an ad hoc networking environment. In *IEEE INFOCOM*, 2001.
13. M. Hajiaghayi, N. Immorlica, and V. S. Mirrokni. Power optimization in fault-tolerant topology control algorithms for wireless multi-hop networks. In *Proceedings of the 9th annual international conference on Mobile computing and networking*, pages 300–312. ACM Press, 2003.
14. R. Jain, A. Puri, and R. Sengupta. Geographical routing using partial information for wireless ad hoc networks. *IEEE Personal Communication*, pages 48–57, February 2001.

15. D. B. Johnson and D. A. Maltz. Dynamic source routing in ad hoc wireless networks. In Imielinski and Korth, editors, *Mobile Computing*, volume 353. Kluwer Academic Publishers, 1996.
16. B. Karp and H. T. Kung. GPRS: Greedy perimeter stateless routing for wireless networks. In *ACM/IEEE International Conference on Mobile Computing and Networking*, 2000.
17. X.-Y. Li, P.-J. Wan, Y. Wang, and C.-W. Yi. Fault tolerant deployment and topology control in wireless networks. In *Proceedings of the 4th ACM international symposium on Mobile ad hoc networking & computing*, pages 117–128. ACM Press, 2003.
18. C. Perkins. Ad-hoc on-demand distance vector routing, 1997.
19. C. Perkins and P. Bhagwat. Highly dynamic destination-sequenced distance-vector routing (DSDV) for mobile computers. In *ACM SIGCOMM'94 Conference on Communications Architectures, Protocols and Applications*, pages 234–244, 1994.
20. V. Raghunathan, C. Schurgers, S. Park, and M. B. Srivastava. Energy-aware wireless microsensor networks. *IEEE Signal Processing Magazine*, pages 40–50, March 2002.
21. V. Rodoplu and T. Meng. Minimum energy mobile wireless networks. In *1998 IEEE International Conference on Communications, ICC'98*, volume 3, pages 1633–1639, Atlanta, GA, June 1998.
22. S. Singh and C. Raghavendra. Pamas: Power aware multi-access protocol with signalling for ad hoc networks. *ACM Computer Communication Review*, July 1998.
23. M. Stemm and R. H. Katz. Measuring and reducing energy consumption of network interfaces in hand-held devices. *IEICE Transactions on Communications*, E80-B(8):1125–31, 1997.
24. I. Stojmenovic. Position-based routing in ad hoc networks. *IEEE Communications Magazine*, July 2002.
25. I. Stojmenovic and X. Lin. Power-aware localized routing in wireless networks. *IEEE Transactions on Parallel and Distributed Systems*, 12(11):1122–1133, 2001.
26. X. Wang, G. Xing, Y. Zhang, C. Lu, R. Pless, and C. Gill. Integrated coverage and connectivity configuration in wireless sensor networks. In *SenSys '03: Proceedings of the 1st international conference on Embedded networked sensor systems*, pages 28–39, New York, NY, USA, 2003. ACM Press.
27. Y. Wang and X.-Y. Li. Geometric spanners for wireless ad hoc networks. In *The 22nd IEEE International Conference on Distributed Computing Systems*, 2002.
28. J. Wu, B. Wu, and I. Stojmenovic. Power-aware broadcasting and activity scheduling in ad hoc wireless networks using connected dominating sets. *Wireless Communications and Mobile Computing*, 4(1):425–438, June 2003.
29. Y. Xu, S. Bien, Y. Mori, J. Heidemann, and D. Estrin. Topology control protocols to conserve energy inwireless ad hoc networks. Technical Report 6, University of California, Los Angeles, Center for Embedded Networked Computing, January 2003. submitted for publication.
30. Y. Xu, J. S. Heidemann, and D. Estrin. Geography-informed energy conservation for ad hoc routing. In *Mobile Computing and Networking*, pages 70–84, 2001.

Adapting Failure Detectors to Communication Network Load Fluctuations Using SNMP and Artificial Neural Nets

Fábio Lima* and Raimundo Macêdo

Distributed Systems Laboratory – LaSiD,
Computing Science Department, Federal University of Bahia,
Campus de Ondina, CEP: 40170-110, Salvador-BA, Brazil
{framon, macedo}@ufba.br

Abstract. A failure detector is an important building block for fault-tolerant distributed computing: mechanisms such as distributed consensus and group communication rely on the information provided by failure detectors in order to make progress and terminate. As such, erroneous information provided by the failure detector (or the absence of it) may delay decision-making or lead the upper-layer fault-tolerant mechanism to take incorrect decisions (e.g., the exclution of a correct process from a group membership). On the other hand, the implementation of failure detectors that can precisely identify failures is restricted by the actual behaviour of a system, especially in settings where message transmission delays and system loads can vary over time. In this paper we explore the use of artificial neural networks in order to implement failure detectors that are dynamically adapted to the current communication load conditions. The training patterns used to feed the neural network were obtained by using Simple Network Management Protocol (SNMP) agents over MIB – Management Information Base variables. The output of such neural network is an estimation for the arrival time for the failure detector to receive the next heartbeat message from a remote process. The suggested approach was fully implemented and tested over a set of GNU/Linux networked workstations. In order to analyze the efficiency of our approach, we have run a series of experiments where network loads were varied randomly, and we measured several QoS parameters, comparing our detector against known implementations. The performance data collected indicate that neural networks and MIB variables can indeed be combined to improve the QoS of failure detectors.

1 Introduction

A distributed system is defined as a collection of processes running on a set of networked, possibly geographically spread, computers. Nowadays, mainly after the widespread use of the World Wide Web, the dependence of society on such systems have become commonplace. This reality has pushed researchers to find

---

* Master Student of the Mechatronics Program at UFBA.

C.A. Maziero et al. (Eds.): LADC 2005, LNCS 3747, pp. 191–205, 2005.

out techniques for building reliable distributed systems, which can deliver the specified services despite failures of some of its components. In particular, distributed consensus has been considered by many researchers as the basic building block to construct other fault-tolerant mechanisms such as non-blocking atomic commitment protocols, replication, and even group membership [1,2]. However, implementing such mechanisms on settings where message transmission delay and processing times cannot be known and bounded (the so-called time-free or asynchronous systems), it is not an easy task. As a matter of fact, it has been proved that consensus is not solvable in such systems when failures may occur [3]. Therefore, no deterministic consensus based fault-tolerant mechanism can be implemented in such scenarios if no extra assumption is considered. That is why many researchers progressively started adopting alternative system models to solve fault-tolerant problems [4,1,5,6]. One of the most referred of such models is the asynchronous system model augmented with unreliable failure detectors proposed by Chandra and Toueg [1,7]. They defined failure detectors in terms of axiomatic properties and proved that consensus problem is solved for some classes of such failure detectors. Nonetheless, in practice, due to variations in the communication transfer delays and in the execution speed of processes in some typical settings (such as computers equipped with time-sharing operating systems connected by a shared Ethernet), the difficulty in implementing failure detectors remains. Even if the required properties are eventually satisfied in a partially synchronous system (and, accordingly, consensus eventually solvable), the uncertainties of unreliable failure detectors may have unacceptable impact on real applications: for instance, consensus decisions may be postponed for an arbitrary amount of time.

Consequently, with the aim of improving the *QoS* of failure detectors, many authors have proposed different techniques to dynamically estimate the timeout values used in failure detection [8,9,10,11,12,13], and Chen, Toueg, and Aguilera proposed a set of metrics that can be used to specify and measure the related *Quality of Service* [14]. In another work, Bertier, Marin, and Sens showed how to build a hierarchical failure detector service that can be customized to distinct applications [15].

Different from existing work, we propose an implementation of failure detectors based on a neural network [1]. Neural networks have successfully been used to pattern recognition in a variety of applications [17]. They possess an adaptive feature that allows each cell within the neural network to modify its state in response to experience. Thus, the neural network can learn and mimic the actual behaviour of a system. In few words, a neural network is an interconnected assembly of simple processing elements. The processing ability of the network is stored in the inter-unit connection strengths, or weights, obtained by a process of adaptation to, or learning from, a set of training patterns. In our work, the training patterns used to feed the neural network were obtained by SNMP (Simple Network Management Protocol) agents over a set of data from a local area

---

[1] A summary and preliminary version of this paper (4 pages) appeared in the proceedings the Brazilian Symposium on Computer Networks [16].

network MIB – Management Information Base [18]. We chose SNMP because it has become the *de facto* standard for network management [19], and therefore, our implementation can be used and interoperate in a variety of system settings. The output of our neural network is an estimation for the time for the failure detector to receive the next *heartbeat* message from a remote process. Using neural networks over MIB data renders our approach a number of advantages. First, as we are not relying on previous received heartbeat messages in order to estimate the arrival time of an expected heartbeat message, we are able to tolerate message omission failures. Second, as an estimation is a function of a given MIB pattern, this solution scales well in terms of the number of monitored processes (i.e., the communication pattern given in terms of MIB variables is not a direct function of the number of monitored processes).

The proposed failure detector was fully implemented and tested over a set of GNU/Linux networked workstations. In order to analyze the efficiency of our approach, we have run a number of experiments where network loads were varied randomly, and we measured several *QoS* parameters, comparing our detector against known implementations. The experiments show that our failure detector performed well when compared with such existing approaches.

The remaining of the paper is organized as follows. In Sect. 2 it is discussed related work. In Sect. 3 it is presented the neural network based failure detector and in Sect. 4 it is analyzed its performance. Finally, in Sect. 5 some conclusions are drawn.

2   Related Work

Failure detection is an important issue for the design of dependable distributed systems. In particular, in the partial synchronous model of the unreliable failure detectors introduced by Chandra and Toueg [1,7], it was presented for the first time a formal definition of failure detectors and related conditions for solving fundamental problems of fault tolerant computing in such a model.

Due to load variations, both in the communication links and in the runtime system, a failure detector can be too *slow*, that is, it may take too much time to suspect a crashed process, and it can make *mistakes* by erroneously suspecting some process that is actually operational. As failure detectors are unreliable, a failure detector may "change its mind" by stopping suspecting a process from which a new message has been received. Thus, a failure detector working on behalf of a process p may alternate by suspecting and trusting a remote process q for an arbitrary period of time. To be useful, however, a failure detector has to be reasonably *accurate* (i.e., must avoid wrong suspicions) and *fast* (i.e., must avoid unnecessary delays in suspecting a failures). In order to evaluate how fast and accurate a failure detector is, as mentioned before, Chen, Toueg, and Aguilera proposed a set of metrics for the *QoS* specification [14]. The aim of these metrics is to describe the failure detector's *speed* (how fast it detects crashes) and its *accuracy* (how well it avoids mistakes). They defined the following three basic metrics and showed how to fully qualify the service of failure detectors based on these metrics.

1. **Detection Time:** defines the failure detector's speed, which is the time that elapses from the moment when a process p crashes to the time when the failure detector starts suspecting p permanently.
2. **Mistake Recurrence Time:** defines the time between two consecutive mistakes. That is, the time that elapses from the moment when a process is erroneously suspected to next time it is again erroneously suspected.
3. **Mistake Duration:** defines the time it takes the failure detector to correct a mistake. That is, the time that elapses from the moment a process is erroneously suspected to the time this suspicion is removed (it stops suspecting the process).

With the aim of improving the *QoS* of failure detectors, many authors have proposed different techniques to dynamically estimate the timeout values used in failure detection. These techniques are either based on the probabilistic behaviour of the system [14] or on monitoring data for control message transmission delays (*heartbeat* or *I-am-alive* messages) [8,9,13], or even a combination of both [10]. The main drawback of some of these approaches lies in the fact that it is not always possible to find out a probabilistic distribution that captures the actual behaviour of the network, especially when communication and load patterns can change in a random fashion. In most of these works, the authors have not measured the *QoS* of their implementations using the metrics formerly defined for the failure detectors in [14].

To the best of our knowledge, two previous works explicitly provided implementations and related *QoS* performance analyzes for their failure detectors. Chen, Toueg, and Aguilera [14], presented implementations for failure detectors that rely on clock synchronization and a probabilistic behaviour of the system. Bertier, Marin, and Sens [10] extended the failure detector developed by Chen, Toueg, and Aguilera, by introducing a safety margin dynamically calculated according Jacobson's algorithm [20], which resulted in a detector with a better detection time average (though, slightly sacrificing the quality of other metrics). None of the existing work has explored the use of neural networks to estimate the arrival time of heartbeat messages.

3   The Neural Network Based Adaptive Failure Detector (NN-AFD)

3.1   System Model and Assumptions

It is assumed a distributed system of processes that are able to communicate with each other through channels that deliver uncorrupted messages. Messages may be lost from time to time, but if an infinite amount of messages m is transmitted, at least one message m will reach its destination (fair loss assumption, or quasi-reliable assumption as in [15]). Processes fail only by crashing (halting execution) without producing any further actions. We consider the model of partial synchrony proposed by Chandra and Toueg in [1], which defines that, in every execution there are bounds on process speeds and on message transmission times, but these bounds are not known and they hold only after some unknown time (the Global Stabilization Time – GST).

The algorithm presented in this paper is based on the push model. That is, it is assumed that all processes will be permanently sending heartbeat messages with a time interval, named HP (heartbeat period), between the emissions of two consecutive heartbeat messages. To monitor a process q, a process p uses an estimated value, named TO (timeout), which tells p how much time it has to wait for the next heartbeat message from q. Then, if after TO p does not receive the next heartbeat message from q, it suspects that q has crashed. In order to be adaptive to the actual communication loads, we allow TO values to vary over time. TO is composed by the sum of two values, the estimated time for the arrival of the next heartbeat message (EA) and the safety margin α. The safety margin is added to avoid false detections.

3.2   The Implementation of the NN-AFD

Our failure detector is adaptive with respect to the current communication load, and we use a neural network to achieve the required adaptation. As mentioned earlier, a neural network is an interconnected assembly of simple processing elements. The basic unit of a neural network represents a neuron. A neuron receives a set of signals from the neurons connected to it and produces an output response. In our work, we use a special kind of neural network called Feedforward Multilayer Perceptron (MLP). In the MLP Neural Network, the neurons are grouped in layers and they are interconnected in the following way: a neuron is connected through links to all neurons of previous and next layers (if any); the neuron of a layer is not linked with another neuron of the same layer; there are three basic type of layers: input, middle and output; the input layer receives the input signal and the number of neurons in it corresponds to the number of data attributes; middle layers interconnect the input layer with the output layer; the quantity of layers and the number of neurons for each layer are unknown and depend on the application; finally, the output layer shows the result of processing. The feedforward means that the amount of data flows in one direction only, from the input to the output.

In order to define the configuration of the neural network and the list of MIB variables to be used, we tried distinct neural network configurations (from simple ones, with less layers and neurons, to more complex ones) and also distinct sets of MIB variables (from smaller to larger sets), measuring always the level of adaptation achieved (reflected in the related failure detection QoS). The final configuration was then chosen when the adaptation observed remained virtually the same as in the previous configuration.

The neural network we implemented has four layers: an input layer with six neurons corresponding collected data from the MIB – Management Information Base [18]; two middle layers, one layer with nine and the other with four neurons, respectively; and an output layer with a single neuron, the estimated timeout for the arrival of the next heartbeat message. This neural network, which was fully implemented using the Java – GNU/Linux environment, is depicted in Fig. 1.

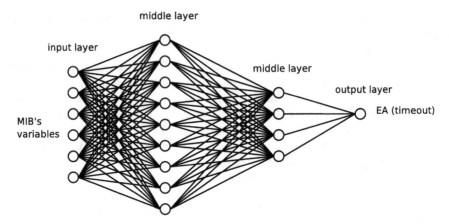

Fig. 1. The four-layer neural network for estimating heartbeat timeouts

3.3   The Use of SNMP to Monitor Communication Loads

Nowadays, computer networks have become complex systems, involving software and hardware components from a variety of vendors. To interoperate, the management protocols used to monitor such systems must comply with some standard. SNMP, an open framework developed by the TCP/IP community, is the *de facto* standard for network management protocols, as it has been adopted by many organizations. SNMP is based on the manager/agent paradigm where management applications (the managers) monitor network devices (or simply components) through agents. Communication between the managers and agents is carried out by the SNMP protocol that runs on top of UDP (User Datagram Protocol). Each agent replies to SNMP queries according to the MIB maintained by the agent (A given MIB holds a collection of objects, each one representing a characteristic of the related managed component). In order to model the system and communication loads we used the six MIB variables [18] listed below.

1. `IfInUcastPkts` - the number of subnetwork-unicast packets delivered to a higher-layer protocol;
2. `ifOutUcastPkts` - the total number of packets that higher-level protocols requested be transmitted to a subnetwork-unicast address, including those that were discarded or not sent;
3. `ifOutQLen` - the length of the output packet queue (in packets);
4. `udpInDatagrams` - the total number of UDP datagrams delivered to UDP users;
5. `udpOutDatagrams` - the total number of UDP datagrams sent from this entity; and,
6. `udpNoPorts` - the total number of received UDP datagrams for which there was no application at the destination port.

3.4 Training and Querying the Failure Detector

Adaptation of NN-AFD is achieved by using the neural network in two distinct phases, namely the training and estimation phases. During the first phase, the Neural Network is trained through the backpropagation algorithm [17][2] to associate patterns of communication loads (represented by MIB patterns) with the timeout value for the arrival of the next heartbeat message (for a fixed frequency for sending heartbeats).

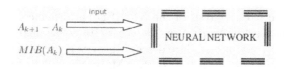

Fig. 2. Training the failure detector

Fig. 3. Querying the failure detector

The pairs communication load/timeout were collected in a real experiment where processes exchanged a number of heartbeat messages, as follows. When a heartbeat message arrives, say hb_k, the current MIB pattern[3] is read and stored together with the arrival time for hb_k say A_k. Afterwards, when the next heartbeat hb_{k+1} arrives, the timeout is calculated by the difference between the arrival time of hb_{k+1}, A_{k+1}, and the stored arrival time of hb_k, A_k. Then, the pair pattern/timeout is built into the neural network. In this phase, we run 19 experiment rounds and for each round we collected pattern/timeout values for 1000 heartbeat messages received (it took approximately 5 hours or 19 x 1000 seconds). At this point we observed that the training phase could be stopped as the level of adaptation observed (i.e., the QoS of the failure detector) remained virtually the same as in the previous round.

During the second phase, the neural network is queried to estimate the arrival time of the next heartbeat message for a particular pattern of the MIB. In both phases, the same time interval between heartbeat messages is used (the heartbeat period, HP). Below it is illustrated the behaviour of the neural network, related

---

[2] In such an algorithm, the input is presented to the network, the calculated output is compared with the expected result and the calculated error is back propagated until the input layer, changing the weight of each link to adjust the network.

[3] Actually, we only considered pattern modifications since the reception of the previous heartbeat.

to the training (Fig. 2) and estimation (Fig. 3) phases, respectively. The time A_k in Fig. 3 refers to the time instant when the network is queried to produce the estimation for the timeout of the next heartbeat and δ is the time spent by the trained network to produce the estimation.

4   The QoS Analysis of NN-AFD

4.1   The Failure Detector Implementations

To allow a comparative analysis of the NN-AFD performance, we implemented in our system the Bertier-Marin-Sens' detector [10], which is, to the best of our knowledge, the only adaptive failure detector previously published and analyzed under the same metrics we utilise. Bertier-Marin-Sens' detector implements a push style adaptive failure detector and through experiments they showed that their failure detector produced, in some circumstances, better QoS compared with the Chen-Toueg-Aguilera's detector [14,11]. However, as the safety margin used in Chen-Toueg-Aguilera's detector is constant during an execution, calculated according to a specified application QoS requirement, comparing both detectors without considering distinct safety margins is meaningless. Hence, we first carry out a thoroughly comparison with Bertier-Marin-Sens' detector. After that, we fix a given safety margin for the Chen-Toueg-Aguilera's detector and show some data comparing the three approaches altogether. Before describing the experiments and data analysis, let us present the Bertier-Marin-Sens' detector we implemented.

Bertier-Marin-Sens' Detector. This detector uses the heartbeat strategy, defined by the heartbeat period, HP, and the timeout delay, TO, to receive the next heartbeat message. In case TO expires, the sending process is suspected of crash until a heartbeat message is received from it.

Bertier-Marin-Sens' detector is a combination of Chen-Toueg-Aguilera's detector and the Jacobson's estimation. Jacobson's estimation has been used in the protocol TCP to estimate the delay after which a node retransmits its last message [20]. To calculate the estimation for the next heartbeat message arrival time, named $TO_{(k+1)}$, the Bertier-Marin-Sens' estimation is calculated by adding the Chen-Toueg-Aguilera's estimation $EA_{(k+1)}$, to the safety margin given by the Jacobson's estimation, $\alpha_{(k+1)}$. That is

$$TO_{(k+1)} = EA_{(k+1)} + \alpha_{(k+1)} \tag{1}$$

The details of both formulas are given below.

Chen-Toueg-Aguilera's Estimation. This implementation estimates the time for the arrival of the next heartbeat message (EA) based on the reception time of the last n heartbeat messages. The estimation for the arrival time of the next heartbeat message ($EA_{(k+1)}$) is given below, where A_i is the arrival time for the ith heartbeat and HP is the time period between the emissions of two heartbeats:

$$EA_{(k+1)} \approx \frac{1}{n} \left(\sum_{i=k-n}^{k} A_i - HP \cdot i \right) + (k+1) \cdot HP \qquad (2)$$

Jacobson's Estimation. In Jacobson's algorithm, the estimation presumes that the behaviour of the system is not constant. Thus, it adapts the safety margin each time it receives a message. The adaptation of the safety margin α uses the error in the last estimation. The parameter γ represents the importance of the new measure with respect to the previous ones. *delay* represents the estimate margin, and *var* estimates the magnitude between errors. β and ϕ permit to ponder the variance; typical values are $\beta = 1$ and $\phi = 4$.

The original algorithm is:

$$error_{(k)} = A_k - EA_{(k)} - delay_{(k)} \qquad (3)$$

$$delay_{(k+1)} = delay_{(k)} + \gamma \cdot error_{(k)} \qquad (4)$$

$$var_{(k+1)} = var_{(k)} + \gamma \cdot \left(|error_{(k)}| - var_{(k)} \right) \qquad (5)$$

$$\alpha_{(k+1)} = \beta \cdot delay_{(k+1)} + \phi \cdot var_{(k+1)} \qquad (6)$$

The NN-AFD Detector. We have implemented two versions of our failure detector. The first version is purely based on the neural network described in Sect. 3. In order to evaluate the application of a safety margin together with the neural network, we also measured the its QoS using a constant safety margin α, varied α from 0 to $2ms$ ($\alpha = 0$ means no safety margin). Thus, after receiving the kth heartbeat hb_k, the estimation for the timeout $TO_{(k+1)}$ meant for the arrival of the heartbeat hb_{k+1}, is given by:

$$TO_{(k+1)} = EA_{(k+1)} + A_k + \alpha \quad \text{(the Pure NN-AFD, see figures 2 and 3)} \qquad (7)$$

Based on the observation that the pure NN-AFD performed better than the Bertier-Marin-Sens' detector when the network load varied randomly (as we show subsequently in our experiments), and slightly worse in stable periods (namely, in the detection time metric), we designed a second version of our failure detector, combining the pure NN-AFD with the Bertier-Marin-Sens' detector, switching between both detectors depending on the network load pattern variation observed. The key point was then to establish the conditions upon which to switch the detectors. For this we calculated the real arrival time for the last three heartbeat messages, say A_k, A_{k-1} and A_{k-2} and calculated whether the sum of the time differences for consecutive messages was larger than a given threshold, defined by us as $\mu + HP \cdot 2$, as below:

if $((A_k - A_{k-1}) + (A_{k-1} - A_{k-2})) > \mu + HP \cdot 2$ then
    use NN-AFD, as in (7)
else
    use Bertier-Marin-Sens' Estimation, as in (1)

The rationale behind the above threshold is to capture a variation on the network load that would delay the last two heartbeat messages for more than μ. Note that in perfect conditions,

$$((A_k - A_{k-1}) + (A_{k-1} - A_{k-2})) = HP \cdot 2 \tag{8}$$

In the experiments we set μ to 10.

4.2   The Experiments

After the NN-AFD detector has been appropriately trained as described in Sect. 3.4, we run a series of experiments in order to evaluate the QoS obtained. The experiments were performed at the Distributed Systems Laboratory (LaSiD) over three Pentium-III 800 Mhz hosts connected by an Ethernet 10/100 Base-T network. The Conectiva Linux 8.0 Operational System with SNMP service and Java (j2sdk 1.4) was used. The network and the hosts were not dedicated to our experiments. That is, we allowed the other users and services to make use of both the hosts (including the three mentioned above) and network during the experiments.

We carried out two kinds of experiments: one with the ordinary network load and another one with an extra load introduced randomly by a third process, named the overloader. As the Bertier-Marin-Sens' detector uses the n most recent heartbeat messages to calculate the $(n + 1)^{th}$ estimation, we first run the system until n ($n = 1000$) messages were received without calculating the QoS information (it took approximately $16.7\,min$). Afterwards, the system starts collecting the necessary information to calculate the QoS performance. The dynamic estimation of the Bertier-Marin-Sens' detector was parameterized, in all experiments, as given in [10]. That is, $\gamma = 0.1$, $\beta = 1$, $\phi = 2$, and $n = 1000$. The variables $error$, $delay$, var, and α had their initial values set to 0.

We also calculated the average time and related standard deviation for the neural network to produce estimations (δ) for all experiments, including, within this measure, the access to the MIB. These figures were $2.141\,ms$ and $0.062\,ms$, respectively. Considering only the calculations realized by the neural network (without the access to the MIB), the average time was $0.053\,ms$ with standard deviation of $0.005\,ms$.

We have chosen to transmit heartbeat messages every second (i.e. $HP = 1000ms$). In the experiments showed in [10], HP was set to $5000\,ms$. By choosing a five time smaller heartbeat transmission period, we are closer to a more realistic model for most applications (as the detection time is necessarily larger than the heartbeat transmission period).

The Experiment with Varied Communication Loads. the overloader process continuously transmitted messages. In the remaining 50 seconds of every

Table 1. QoS performance with random load variation

	Pure NN-AFD	Bertier-Marin-Sens	Combined NN-AFD
False detections	14.17	17.00	16.33
Mistake duration $(ms)^*$	26.05	31.85	31.87
	22.25	24.92	24.17
	24.41	19.34	22.21
Detection time $(ms)^*$	1010.03	1032.17	1010.45
	1006.33	1003.67	1004.00
	17.87	33.22	20.18

* average / median / standard deviation

minute, the overloader process transmits no messages. For each host, there were two overloader processes (in total, six of them: two in the host transmitting heartbeats, two in the host executing the failure detectors, and two in the other host used in the experiments).

These experiments were carried out for six rounds. In each round, after sending 1000 heartbeat messages as required by the adaptation phase of the Bertier-Marin-Sens' detector, a process in a given host, say p, sent heartbeat messages for about 10 minutes, with the frequency of 1 message per second ($HP = 1000\ ms$). Then, a process in another host tries to detect the failure of p by running in parallel the three detectors: the Pure NN-AFD, the combined NN-AFD, and the Bertier-Marin-Sens' detector.

The data collected in each round included the arrival time for the heartbeat messages, according to local clock, the estimated heartbeat arrival time for each detector, the time that a detector starts suspecting p and the time that a detector corrects a mistake (i.e., stops suspecting p as a message from p arrives). We then calculated for each round, the number of false detections, the average, median, and standard deviation for the detection time and mistake duration. Table 1 shows the results. The mistake duration, the detection time, and the number of false detections are the average for the six rounds.

We also run the experiments above for distinct safety margins applied to the NN-AFD and combined NN-AFD (that is, for $\alpha = 1$ and $\alpha = 2$). The performance data showed that the average detection time was worse in both experiments (which is expected), without improvements regarding the number of false detections and average mistake duration.

Experiment with Ordinary Communication Load. This experiment is similar to the previous one. The only difference is that the overloader processes were not running. The aim of this scenario is to compare the performance of the failure detectors in a stable communication link.

The Experiment with the Chen-Toueg-Aguilera's Detector. In order to evaluate the influence of the use of a safety margin within the Chen-Toueg-

Aguilera's detector, we carried out another experiment where the three detectors were run together with ordinary communication load. For the Chen-Toueg-Aguilera's detector, we defined a constant safety margin of 30 milliseconds and 500 heartbeat messages were transmitted. Table 3 show the data colleted comparing the four detectors.

Discussion. When we consider a network with load variations (Tab. 1), our experiments clearly show that the pure NN-AFD performed better than the detector of Bertier-Marin-Sens for all metrics. Indeed, when compared with Bertier-Marin-Sens' detector, this result contradicts the common intuition that shorter detection time, which favours faster recovery procedures for fault-tolerant computing, leads to a less accurate detector.

Still considering the experiment with load variation, the combined NN-AFD outperformed Bertier-Marin-Sens' detector, being only slightly worse in the mistake duration metric (in fact, practically the same values). As expected, NN-AFD slightly outperformed the combined NN-AFD. Indeed, these results validate the technique used to switch between the detectors.

As for the experiments with ordinary loads, the data in Tab. 2 show that the NN-AFD detector is more accurate than the Bertier-Marin-Sens' detector as it presents a better mistake duration and a smaller number of false suspicions. However, it had a slightly worse performance regarding the third metric (detection time). On the other hand, the combined NN-AFD had a very similar performance to the Bertier-Marin-Sens' detector, which is an indication that the technique used to switch between the detectors properly detected stable load patterns, switching to the Bertier-Marin-Sens' detector properly.

In order to compare our results with the detector of Chen-Toueg-Aguilera, let us recall the figures published in [10] comparing Bertier-Marin-Sens' and Chen-Toueg-Aguilera's detectors, for $HP = 5000\,ms$, in two experiments (one with constant load and another with ordinary load). In both experiments, Bertier-Marin-Sens' detector performed better in terms of detection time, but it was less accurate regarding the other metrics. We have confirmed this result with the experiment showed in Tab. 3. Notice that in this particular experiment, the NN-AFD detector

Table 2. QoS performance with ordinary load

	Pure NN-AFD	Bertier-Marin-Sens	Combined NN-AFD
False detections	8.50	9.00	9.50
	25.36	27.33	26.82
Mistake duration $(ms)^*$	16.50	18.50	17.50
	22.25	22.35	21.85
	1005.23	1003.80	1003.71
Detection time $(ms)^*$	1005.00	1002.50	1002.50
	0.66	6.22	3.75

* average / median / standard deviation

Table 3. Performance with constant safety margin

	Pure NN-AFD	Bertier-Marin-Sens	Combined NN-AFD	Chen-Toueg-Aguilera
False detections	3	6	6	–
Mistake duration $(ms)^*$	17.33	17.83	17.83	–
	21.00	20.50	20.50	–
	3.86	3.89	3.89	–
Detection time $(ms)^*$	1010.72	1003.21	1003.66	1031.69
	1007.00	1002.00	1002.00	1032.00
	15.03	6.68	5.22	5.74

* average / median / standard deviation

was more accurate than the Bertier-Marin-Sens' detector (less false detections), with a worse detection time (consistent with the data showed in Tab. 2). The high detection time of the Chen-Toueg-Aguilera's detector was due to the safety margin of 30 milliseconds. By choosing smaller safety-margins, this average will decrease accordingly. It is left open, however, how smaller this safety-margin can go without affecting the QoS of the other metrics.

A justification to choose the Bertier-Marin-Sens' detector, instead the Chen-Toueg-Aguilera's detector, it is to achieve a better detection time (as showed in [10]). On the other hand, our experiments show that even in the situations where the Bertier-Marin-Sens' detector performed better than NN-AFD (Tab. 2), the combined NN-AFD had virtually the same QoS as Bertier-Marin-Sens' detector. Therefore, by using the combined NN-AFD and choosing the proper technique to switch between the detectors, one can take full advantage of the good characteristics of NN-AFD, which clearly has the best performance in situations where load patterns very randomly, still keeping the good QoS of the Bertier-Marin-Sens' detector in stable periods.

5   Concluding Remarks

Being able to detect failures is a fundamental issue in designing fault-tolerant distributed systems. However, the actual behaviour of a distributed system limits the ability of providing such a mechanism. Whereas synchronous systems allow for the construction of perfect failure detection based simply on fixed timeouts, accurate failure detection cannot be achieved for fully asynchronous systems or even partially synchronous systems, as it is the case for the partial synchronous model of unreliable failure detectors of Chandra and Toueg [1,7].

To precisely define the failure detector QoS requirements, Chen, Toueg, and Aguilera, proposed a set of metrics and showed an implementation of an adaptive algorithm that performed well under the given metrics. Bertier, Marin, and Sens, extended the failure detector developed by Chen, Toueg, and Aguilera, by introducing a safety margin dynamically calculated according Jacobson's algorithm [20], which resulted in a detector with a better detection time average.

In the present paper we explored the use of artificial neural networks in order to improve the quality of service of failure detectors. The training patterns used to feed the neural network were obtained by using SNMP agents over MIB data related to a local area network. The output of the neural network is an estimation for the arrival time for the failure detector to receive the next heartbeat message from a remote process. The failure detector was fully implemented and tested over a set of GNU/Linux networked workstations. In order to analyze the efficiency of our approach, we have run a series of experiments where network loads were varied randomly, and we measured several QoS parameters according to the metrics introduced in [14], comparing the introduced detector against known implementations. The experiments show that the presented detector performed well compared with existing approaches and, therefore, it is an indication that neural networks and MIB variables can be combined together to improve the QoS of failure detectors. Whereas the specific neural network chosen and the related MIB variables yield a good QoS performance, it is left open, however, how much improvement can be archived considering distinct types of neural networks with distinct combinations of MIB variables.

In the next step of our work, we are going to develop an adaptation layer, also based on a neural network, which will take distinct application failure detection QoS requirements and produce the adequate heartbeat period for given system settings. Moreover, we will explore MIB data collected from remote network segments. This will allow us to use our approach in a wider variety of application scenarios, including the Internet.

Acknowledgments

The authors would like to thank the anonymous referees whose comments helped us to improve the paper presentation, and the financial support of SANMINA-SCI (project 'Service Integration and Failure Monitoring' a collaboration project between LaSiD/DCC/Federal University of Bahia and SANMINA-SCI).

References

1. Chandra, T.D., Toueg, S.: Unreliable failure detectors for reliable distributed systems. Journal of the ACM **43** (1996) 225–267
2. Hurfin, M., Macêdo, R., Raynal, M., Tronel, F.: A general framework to solve agreement problems. In: SRDS '99: Proceedings of the 18th IEEE Symposium on Reliable Distributed Systems, Washington, DC, USA, IEEE Computer Society (1999) 56–65
3. Fischer, M.J., Lynch, N.A., Paterson, M.S.: Impossibility of distributed consensus with one faulty process. Journal of the ACM **32** (1985) 374–382
4. Dwork, C., Lynch, N., Stockmeyer, L.: Consensus in the presence of partial synchrony. Journal of the ACM **35** (1988) 288–323
5. Verissimo, P., Casimiro, A., Fetzer, C.: The timely computing base: Timely actions in the presence of uncertain timeliness. In: DSN '00: Proceedings of the 2000 International Conference on Dependable Systems and Networks, Washington, DC, USA, IEEE Computer Society (2000) 533–542

6. Gorender, S., Macêdo, R., Raynal, M.: A hubrid and adaptive model for fault-tolerant distribuded computing. In: DSN '05: Proceedings of the 2005 International Conference on Dependable Systems and Networks, Yokohama, Japan, IEEE Computer Society (2005) 412–421

7. Chandra, T.D., Hadzilacos, V., Toueg, S.: The weakest failure detector for solving consensus. Journal of the ACM **43** (1996) 685–722

8. Macêdo, R.: Failure detection in asynchronous distributed systems. In: II WTF: Workshop on Tests and Fault-Tolerance, Curitiba, PR, Brazil (2000) 76–81

9. Nunes, R., Jansch-Porto, I.: A lightweight interface to predict communication delays using time series. In: LADC '03: Proceedings of the Latin-American Symposium on Dependable Computing, São Paulo, SP, Brazil (2003) 254–263

10. Bertier, M., Marin, O., Sens, P.: Implementation and performance evaluation of an adaptable failure detector. In: DSN '02: Proceedings of the 2002 International Conference on Dependable Systems and Networks, Washington, DC, USA, IEEE Computer Society (2002) 354–363

11. Devianov, B., Toueg, S.: Failure detector service for dependable computing. In: DSN '00: Proceedings of the 2000 International Conference on Dependable Systems and Networks, Washington, DC, USA, IEEE Computer Society (2000) B14–B15

12. Larrea, M., Fernández, A., Arévalo, S.: Optimal implementation of the weakest failure detector for solving consensus (brief announcement). In: PODC '00: Proceedings of the nineteenth annual ACM symposium on Principles of distributed computing, New York, NY, USA, ACM Press (2000) 334

13. Sotoma, I., Madeira, E.R.M.: Adaptation - algorithms to adaptive fault monitoring and their implementation on corba. In: DOA '01: Proceedings of the Third International Symposium on Distributed Objects and Applications, Washington, DC, USA, IEEE Computer Society (2001) 219–228

14. Chen, W., Toueg, S., Aguilera, M.K.: On the quality of service of failure detectors. IEEE Trans. Comput. **51** (2002) 13–32

15. Bertier, M., Marin, O., Sens, P.: Performance analysis of hierarchical failure detector. In: DSN' 03: Proceedings of the 2003 International Conference on Dependable Systems and Networks, San-Francisco (USA), IEEE Society Press (2003) 635–644

16. Macêdo, R., Lima, F.: Improving the quality of service of failure detectors with snmp and artificial neural networks. In: Simpósio Brasileiro de Redes de Computadores, SBRC'2004 (short-paper track, Gramado-RS, Brazil, SBC - Brazilian Computer Society (2004) 583–586

17. Haykin, S.: Neural Networks: A Comprehensive Foundation. 1 edn. MacMillan Publishing Company (1994)

18. McCloghrie, K., Rose, M.: Management Information Base for Network Management of TCP/IP-based internets:MIB-II. RFC 1213 (Standard) (1991) Updated by RFCs 2011, 2012, 2013.

19. Case, J., Fedor, M., Schoffstall, M., Davin, J.: Simple Network Management Protocol (SNMP). RFC 1157 (Historic) (1990)

20. Jacobson, V.: Congestion avoidance and control. In: SIGCOMM '88: Symposium proceedings on Communications architectures and protocols, New York, NY, USA, ACM Press (1988) 314–329

Parsimony-Based Approach for Obtaining Resource-Efficient and Trustworthy Execution[*]

HariGovind V. Ramasamy, Adnan Agbaria, and William H. Sanders

Coordinated Science Laboratory, University of Illinois at Urbana-Champaign,
1308 W. Main Street, Urbana, IL 61801, USA
{ramasamy, adnan, whs}@crhc.uiuc.edu

Abstract. We propose a resource-efficient way to execute requests in Byzantine-fault-tolerant replication that is particularly well-suited for services in which request processing is resource-intensive. Previous efforts took a failure-masking *all-active* approach of using all $2t + 1$ execution replicas to execute all requests, where t is the maximum number of failures tolerated. We describe an asynchronous execution protocol that combines failure masking with imperfect failure detection and checkpointing. Our protocol is parsimony-based since it uses only $t + 1$ execution replicas, called the primary committee or pc, to execute the requests normally. Under normal conditions, characterized by a stable network and no misbehavior by pc replicas, our approach enables a trustworthy reply to be obtained with the same latency as in the all-active approach, but with only about half of the overall resource use of the all-active approach. However, a request that exposes faults among the pc replicas will incur a higher latency than the all-active approach mainly due to fault detection latency. Under such conditions, the protocol switches to a recovery mode, in which all $2t + 1$ replicas execute the request and send their replies. Then, after selecting a new pc, the request latency returns to the same level as that of all-active execution. Practical observations point to the fact that failures and instability are the exception rather than the norm. That motivated our decision to optimize resource efficiency for the common case, even if it means paying a slightly higher performance cost during periods of instability.

1 Introduction

The trustworthiness of a networked information system (NIS) is judged by its ability to provide security and fault tolerance despite software errors, operator errors, and malicious attacks [1]. Since it is difficult to constrain the behavior of a compromised node that is under the control of an adversary, the Byzantine failure model is an attractive way to model such behavior. By using redundancy to mask the effects of up to a threshold number of security-compromised or failed nodes, Byzantine fault tolerance (BFT) is a promising approach to enhance

---

[*] This material is based upon work supported in part by the National Science Foundation under Grant No. CNS-0406351 and DARPA contract F30602-00-C-0172.

C.A. Maziero et al. (Eds.): LADC 2005, LNCS 3747, pp. 206–225, 2005.

the trustworthiness of NISs. In BFT replication, the replicas of a service run deterministic state machines [2] and execute client requests in the same order to ensure state consistency. Execution of requests is preceded by an agreement among the replicas on the request-delivery order using Byzantine agreement (or, equivalently, atomic broadcast).

While Byzantine fault tolerance (BFT) as an area has existed for more than two decades, much of the earlier work had significant but mainly theoretical implications. More recent work has focused on removing the barriers that limit the widespread use of BFT to improve security and reliability.

Castro and Liskov's BFT library [3] showed that BFT replication systems can be built that add only modest extra latencies relative to unreplicated systems. They also showed that *proactive recovery* can be used to significantly increase the coverage of the assumption that there are at most a threshold number (one-third) of replicas that can be corrupted by the adversary.

A drawback of BFT replication that limited its applicability in many real-world settings was the requirement that all replicas should run the same service implementation and update their states deterministically. If all replicas ran the same service implementation, then an adversary could exploit the same vulnerabilities or software bugs to cause all replicas to fail simultaneously. The determinism requirement is non-trivial to satisfy in many real-world services. Rodrigues et al. [4] proposed an extension of the BFT library called BASE, which uses abstraction to address that drawback. Specifically, BASE enables the use of diverse COTS-based replica implementations, thereby reducing the possibility of common-mode failures. Their technique uses wrappers to ensure that diverse and non-deterministic implementations of the replicas of a service satisfy a common abstract specification.

Yin et al. [5] improved BASE by enforcing a clean separation between agreement on the request delivery order and execution of requests in the agreed-upon order. Figure 1(b) gives a high-level view of the separation, and contrasts it with traditional BFT (Fig. 1(a)), which tightly couples agreement and execution. While the *agreement cluster* has the usual $3t + 1$ replicas (we call them *agreement replicas*), the separation allowed the number of replicas in the *execution cluster* (we call them *execution replicas*) to be decreased from $3t + 1$ to $2t + 1$, where t is the number of simultaneous replica faults that have to be tolerated. The separation also opened up the possibility of including a privacy firewall between the two phases that could be used to enhance confidentiality by preventing a malicious replica in the execution cluster from disclosing unauthorized information to users.

This paper proposes a resource-efficient way to execute requests in BFT replication that is particularly well-suited for services in which request execution is resource-intensive (e.g., computation-intensive). The previous best way was the one proposed by Yin et al. that used $2t + 1$ execution replicas. Previous work followed an *all-active* approach (Figures 1(a) and (b)), in which all execution replicas executed the request. We observe that while $2t + 1$ execution replicas is the minimum number of replicas needed to mask t corrupt ones, the client

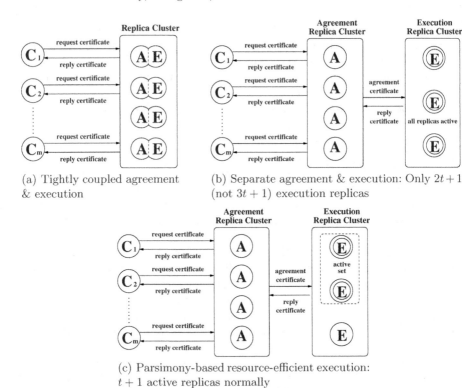

(a) Tightly coupled agreement & execution

(b) Separate agreement & execution: Only $2t + 1$ (not $3t + 1$) execution replicas

(c) Parsimony-based resource-efficient execution: $t + 1$ active replicas normally

Fig. 1. Successive Steps for Obtaining Efficient Execution in BFT Replication

needs only a set of $t + 1$ identical replies (we call this set the *reply certificate*) before considering the reply to be trustworthy. The reason is that identical replies from $t + 1$ execution replicas will always include the reply from at least one correct replica. Hence, that reply value must be correct. We leveraged the above observation and designed an optimistic protocol for the execution cluster replicas. The protocol is parsimony-based since normally only a fraction of the available resources (i.e., $t + 1$ out of $2t + 1$ replicas) are used for request execution.

Our protocol is based on the *optimistic hope* [6] that normally the network is well-behaved and a designated set of $t + 1$ replicas function properly. When the optimistic hope is satisfied, reply certificates are obtained with the same latency, but with only about half of the overall resource use of the all-active approach. *Overall resource use* is the average resource use at a replica times the number of replicas.

The approach does have a price: in situations when the optimistic hope is not satisfied, the latency for obtaining the reply certificate is higher than it is in the all-active approach due to failure-detection latency. However, even under such situations, our protocol guarantees safety and liveness, subject only to the condition that messages are delivered eventually. Even in NISs that are high-value attack targets, such situations are expected to be rare. Hence, it makes sense to optimize for the common case, and be prepared for the rare situations in which a higher price may be paid.

The rest of this paper is organized as follows. Section 2 presents the system model and assumptions. Section 3 describes an abstraction of the agreement cluster that simplifies our protocol presentation. Section 4 presents our parsimony-based execution protocol in detail. Section 5 specifies the properties that our protocol is expected to satisfy. We have implemented our protocol and evaluated its performance through fault injection experiments, the results of which are described in Sect. 6. Section 7 lists some practical applications for our protocol and compares it with related work. Finally, Sect. 8 presents our conclusions.

2   System Model

We consider an asynchronous distributed system model in which nodes may operate at arbitrarily different speeds. Every pair of nodes is connected by a *secure asynchronous channel* that provides authenticity. Authenticity can be easily ensured using cheap message authentication codes (MACs) [7]. Asynchronous channels mean that there are no *a priori* bounds on message transmission delays.

The BFT-replicated service consists of n_a replicas forming an agreement cluster and n_e replicas forming an execution cluster. Agreement cluster replicas and execution cluster replicas may occupy different nodes (i.e., there is a physical separation between agreement and execution) or may share nodes (i.e., there is only logical separation between agreement and execution). Clients and replicas occupy different nodes.

Figure 1 shows the dataflow from the clients to the replicated service and back. Clients send authenticated *request certificates* to the agreement cluster replicas. The request certificates will carry some validating information showing that the clients do have the privilege to issue the requested operations. The agreement cluster replicas run a Byzantine agreement or BFT atomic broadcast protocol (e.g., Castro-Liskov's BFT protocol [3] or Cachin et al.'s atomic broadcast protocol [8]) to agree on the order of request execution. The agreed-upon order is conveyed to the execution cluster replicas through *agreement certificates* that show that a sufficient number of agreement cluster replicas approved the order. The execution cluster replicas start with the same initial service state and implement deterministic state machines; they convey the result of executing the requests through reply certificates that contain evidence showing that the result is indeed correct. The reply certificates are sent to the agreement cluster replicas, which then forward the reply certificates to the client.

A computationally bounded adversary controls up to t agreement cluster replicas and up to t execution cluster replicas. We call the replicas controlled by the adversary *corrupt*; other replicas are *correct*. Corrupt replicas may behave in an arbitrary (i.e., Byzantine) manner. Further, it is well-known that to mask t faults, the minimum number of agreement cluster replicas needed is $3t + 1$ and the minimum number of execution cluster replicas needed is $2t + 1$. Thus,

$n_a \geq 3t + 1$ and $n_e \geq 2t + 1$. Figure 1(c) depicts the situation where $t = 1$, $n_a = 4$, and $n_e = 3$.

The adversary controls the network and determines the scheduling of messages on all the channels. Timeouts are messages a replica sends to itself; hence, the adversary controls the timeouts as well. However, the parsimony-based execution protocol's properties are guaranteed only to the extent that messages exchanged between correct replicas are eventually delivered without any change in contents.

Besides MAC-authentication for implementing secure channels, we also use public-key signatures [7]. A recipient of a signed message that is convinced of the message's authenticity can convince a third party about the message's authenticity. However, MAC-authentication is not provable to a third party. For the public key signature scheme, each replica possesses a public key, private key pair. The public key of a replica is known to all other replicas. We assume that the signature scheme is secure in the sense of the standard security notion for signature schemes of modern cryptography, i.e., existential forgery against chosen-message attacks [9].

3   The Agreement Cluster Abstraction

In the description of the parsimony-based execution protocol, we consider the agreement cluster as an abstract service that guarantees certain properties relating to the ordering of client requests. We use $\mathcal{AC}$ to denote that service. Abstracting the n_a agreement cluster replicas as one logical entity allows us to keep the focus on the execution cluster replicas with whose behavior the parsimony-based execution protocol is concerned. The functionality provided by $\mathcal{AC}$ is the binding of sequence numbers (starting from 1 and without gaps) to request certificates, and the conveying of the bindings to the execution cluster through agreement certificate messages. The $\mathcal{AC}$ does not require any information about what execution cluster replicas constitute the pc and sends the agreement certificate messages to all the replicas. An agreement certificate message binds a sequence number s to a client's request certificate. In our protocol description, the message has the form (agree, $s, o, flag$), where the retransmit $flag$ is either true or false. For notational simplicity, we include only the service operation o contained in the client's request certificate rather than the full certificate. First, $\mathcal{AC}$ sends an agree message with the $flag$ value false. If $\mathcal{AC}$ does not receive a reply certificate before a timeout, then it retransmits the agree message with the $flag$ value true. We use the term first-time agree(s) to denote the agree message with sequence number s and $flag$ value false. We use the term retransmit agree(s) to denote the agree message with sequence number s and $flag$ value true.

The $\mathcal{AC}$ provides the following guarantees to the execution cluster replicas:

Agreement: If a correct execution replica receives an agreement certificate binding sequence number s to request certificate rc, then no other correct execution cluster replica receives an agreement certificate binding s to another request certificate rc', where $rc' \neq rc$.

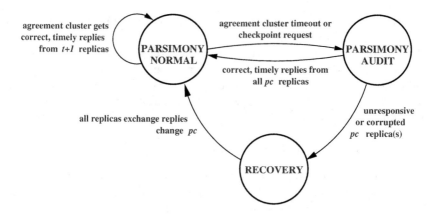

Fig. 2. The Three Modes of Protocol Operation

Liveness: If a client sends a request certificate r to $\mathcal{AC}$, then all correct execution cluster replicas eventually receive an agreement certificate binding some sequence number s to rc.

The above properties of the $\mathcal{AC}$ allow the BFT-replicated service to tolerate an arbitrary number of corrupt clients: even if corrupt clients' requests are executed, those clients cannot cause the service states of correct execution cluster replicas to become inconsistent. Client access control and request-filtering policies [3,5] can be enforced in the implementation of $\mathcal{AC}$; the policies can effectively limit the number and scope of requests from corrupt clients.

4   The Protocol

To generate a reply certificate for one client request, the execution cluster replicas may go through at most three modes of protocol operation, as shown in Fig. 2. We now describe the protocol's operation in each of those three modes and the triggers that cause the transitions among the modes.

In the following description, we use $E_1, E_2, \ldots, E_{n_e}$ to denote the execution cluster replicas. The *rank* of replica E_i is i. $\langle m \rangle_{\sigma_i}$ is used to denote a message m signed by replica E_i. E_i maintains a local sequence number variable s, where $s - 1$ indicates the highest sequence number for which E_i is sure that $\mathcal{AC}$ has obtained a reply certificate.

Each replica maintains two sets, *slow* and *corrupt*, initialized to empty sets. The $(t + 1)$ lowest-ranked replicas that are neither in *slow* nor in *corrupt* constitute what we call a *primary committee*, or *pc* for short. We call the t non-*pc* execution replicas *backups*. Hence, initially, the primary committee at all replicas consists of the $(t + 1)$ lowest-ranked replicas, namely $\{E_1, E_2, \ldots, E_{t+1}\}$. If two replicas have the same *slow* and *corrupt* sets, then their respective primary committees will also be the same. For example, if $t = 3$, then the primary committee at all replicas would initially be $\{E_1, E_2, E_3, E_4\}$. If replica E_2 was later added to replica E_4's *slow* or *corrupt* set, then the primary committee at replica E_4 would become $\{E_1, E_3, E_4, E_5\}$.

To simplify the description of the parsimony-based execution protocol, we assume that $\mathcal{AC}$ sends its next request after receiving the reply certificate for its previous request, i.e., the $\mathcal{AC}$ has only one outstanding request. It is easy to extend the protocol to the case where $\mathcal{AC}$ has any fixed constant number of outstanding requests.

4.1   Parsimony-Based Normal Mode

When E_i receives a *first-time* `agree`$(s+1)$ message, the protocol at E_i moves to the parsimony-based normal mode. E_i maintains a queue of requested operations called *requests*, and adds the service operation indicated in the `agree` message to the queue. Because of our assumption that $\mathcal{AC}$ has at most one outstanding request, E_i can be sure that $\mathcal{AC}$ has obtained a reply certificate for `agree`(s) when it receives the *first-time* `agree`$(s+1)$ message. Hence, E_i increments its sequence number variable s.

If $E_i \notin pc$, then it does nothing more in this mode. On the other hand, if $E_i \in pc$, then it executes the service operation indicated in the `agree` message, and sends the result r of the execution to the $\mathcal{AC}$ in a `reply` message of the form $(\mathtt{reply}, i, s, r)$. E_i also adds its `reply` message to *replies*, a data structure that all replicas have to store `reply` messages from themselves and other replicas. Since the replicated state machines are deterministic and the request execution is done in sequence number order, the r values in the `reply` messages sent by all correct replicas will be identical.

In the normal case, the `reply` messages from the pc replicas will be sufficient for the $\mathcal{AC}$ to obtain a reply certificate, which it then forwards to the respective client. $\mathcal{AC}$ can then issue the `agree` message with the next sequence number $s+1$.

Transition from Normal to Audit Mode. The protocol at E_i transitions to the parsimony-based audit mode from the parsimony-based normal mode when

1. E_i receives a *retransmit* `agree`(s) message from $\mathcal{AC}$ and thereby learns that $\mathcal{AC}$ did not get a reply certificate for `agree`(s) in a timely manner, or
2. E_i receives a *checkpoint request*.

A checkpoint request is a message of the form $(\mathtt{agree}, s+1, o, \mathtt{true})$, where $s+1$ is divisible by the checkpoint interval δ. After every $\delta - 1$ `agree` messages, $\mathcal{AC}$ generates a special `agree` message in which the requested operation o is a *checkpoint* operation[1]. When E_i receives the checkpoint request, E_i knows that $\mathcal{AC}$ must have received a reply certificate for `agree`(s), and hence increments s. Executing a checkpoint operation involves taking a snapshot of the replicated service states and computing the digest of the snapshot. The result field r of the `reply` message for a checkpoint request will contain the checkpoint digest. If E_i

---

[1] Alternatively, execution cluster replicas can *self-issue* a checkpoint request after δ requests from $\mathcal{AC}$.

has obtained a reply certificate for a checkpoint request with sequence number s, we say that the $(s/\delta)^{\text{th}}$ checkpoint is *stable* at E_i. Checkpointing, as will be shown later, is useful for the efficient update of a backup's state when it has to switch to the recovery mode. Checkpointing also allows the garbage collection of `reply` and `agree` messages with sequence numbers less than that of the last stable checkpoint.

4.2   Parsimony-Based Audit Mode

Upon switching to the audit mode, an execution cluster replica E_i starts a timer and expects to obtain a reply certificate before the timer expiry. If progress is not being made, the replicas collectively switch the protocol to the recovery mode, in which all correct replicas generate their own reply messages (if they hadn't done so previously) and ensure that the $\mathcal{AC}$ obtains a reply certificate. If, on the other hand, the replicas indeed receive a reply certificate in a timely manner from the *pc*, they forward the certificate to $\mathcal{AC}$ and the protocol will switch back to the parsimony-based normal mode.

To enable the monitoring of progress, the *pc* replicas are required to send signed `reply` messages to all execution cluster replicas. A *pc* replica E_j retrieves the result value of the `reply` message from the *replies* data structure if it had previously sent a `reply` message to $\mathcal{AC}$ in the parsimony-based normal mode. Otherwise, E_j obtains the result value by executing the operation specified in the corresponding `agree` message from $\mathcal{AC}$.

Transition from Audit to Recovery Mode. An execution cluster replica E_i may not be able to obtain a reply certificate before its local timer expiry for one or both of the following reasons:

1. Slow Replies: A *pc* replica is (deliberately or unintentionally) slow in sending its reply message.
2. Wrong Replies: A *pc* replica did send its reply message, but with the wrong result value.

Slow Replies. If E_i does not receive the `reply` message from a *pc* replica E_k in a timely manner, then E_i sends a signed `suspect` message for E_k to all execution cluster replicas. The `suspect` message has the form $\langle \text{suspect}, i, k, s, c \rangle_{\sigma_i}$, where s is the sequence number and c is a variable called the *reset counter*. The reset counter is an artefact of imperfect failure detection and is used to keep track of the number of times the *slow* set is reset to account for that imperfection. We discuss this in detail in Sect. 4.5.

If a correct replica E_j (has received or) later receives E_k's `reply` message with sequence number s, then E_j simply forwards E_k's `reply` message to E_i upon receiving E_i's `suspect` message. This `reply` forwarding ensures that if at least one correct replica has received E_k's `reply` message for $\text{agree}(s)$, then all correct replicas will eventually receive E_k's message.

On the other hand, if no correct replica has received E_k's `reply` message for $\text{agree}(s)$ in a timely fashion (determined by the replicas' respective local timers), then each of the $n - t$ correct replicas will generate a `suspect` message for E_k. E_i keeps track of all the `suspect` messages it receives by storing them in a data structure, *suspects*.

After receiving $\langle \texttt{suspect}, j, k, s, c \rangle_{\sigma_j}$ messages from $n - t$ distinct E_js (possibly including itself), replica E_i adds E_k to its *slow* set. E_i then sends an `indict` message of the form $(\texttt{indict}, k, s, c, proof)$ to all replicas, where *proof* contains the signed `suspect` messages. E_i also adds s to a set *mustDo* that is used to keep track of the sequence numbers of those requests that caused the protocol to switch to recovery mode; the set is so named because all replicas, whether *pc* or backup, *must* send their own `reply` messages for those requests. Having added the *pc* replica E_k to its *slow* set, E_i updates its *pc* accordingly. E_i then switches to the recovery mode. Any replica E_j at which $E_k \notin slow$ that receives E_i's `indict` message will add E_k to its *slow* set, send its own similar `indict` message for E_k to all replicas, and switch to the recovery mode.

Wrong Replies. Since the state machines are deterministic and request execution is done in sequence number order, any difference in the result values of `reply` messages from two replicas indicates that at least one of them is corrupted. However, to be able to pinpoint in a provable manner which of those two replicas is corrupt, a reply certificate is needed; any replica whose `reply` message contains a result value different from that in a reply certificate is corrupt. Replica E_i sends an `implicate` message of the form $(\texttt{implicate}, s, proof)$ to all replicas, where *proof* contains two or more `reply` messages with differing result values. A recipient E_k of E_i's `implicate` message will not know which of the implicated replicas is actually corrupt, but will be convinced of the need to switch to the recovery phase and add s to the *mustDo* set.

Repeated Transitions from Normal to Audit Mode. A corrupt *pc* replica can cleverly degrade protocol performance by repeatedly refraining from sending a `reply` message to the $\mathcal{AC}$, thereby forcing a transition from the normal to audit mode, while behaving properly in the audit mode. That would result in frequent transitions from the normal to audit mode and back to normal mode, without a change in the *pc*.

The protocol addresses the above problem as follows. If the fraction of requests that resulted in a transition from the normal to audit mode exceeds a fixed threshold, the protocol operates semi-permanently in the audit mode until the next transition to the recovery mode. After the *pc* is changed in the recovery mode, the protocol reverts back to the normal mode.

4.3   Recovery Mode

Only the *pc* replicas send `reply` messages in the normal and audit modes. In the recovery mode, however, backups are also required to send signed `reply` messages to other replicas. Because at least $t + 1$ replicas are correct, the recovery mode guarantees that a reply certificate for `agree`(s) will eventually be obtained. As in the audit mode, the reply certificate is then forwarded to the $\mathcal{AC}$. The execution replicas then change the *pc*, and switch back to the parsimonious normal mode for the next request.

To send a `reply` message, a backup first has to determine the result value corresponding to the request contained in `agree(s)`. As before, the result is obtained from a reply certificate (if previously received), or otherwise by actual execution of the request. Before executing the operation specified in the `agree(s)` message, however, a backup E_i has to ensure that its state is up-to-date. For this purpose, all replicas maintain a variable *updated* to keep track of how up-to-date their state is. Only when *updated* becomes equal to $s - 1$ can E_i execute the operation specified in the `agree(s)` message. Bringing the state up to date may involve two steps:

1. If *updated* < *stable* at E_i, where *stable* is the sequence number of E_i's last stable checkpoint, then E_i first obtains the state corresponding to the execution of all requests with sequence numbers up to *stable*. E_i determines the $t + 1$ replicas whose `reply` messages form the reply certificate for `agree(stable)`. E_i then requests the state corresponding to that checkpoint by sending a message of the form (`state`, *stable*) to those $t + 1$ replicas. Since at least one of the replicas is correct, E_i is guaranteed eventually to obtain the state corresponding to that checkpoint. E_i can easily verify whether the state transferred is correct; E_i computes the digest of a copy of the state obtained after it has applied the updates indicated in the state transfer, and then compares the digest with the one present in the certificate for the stable checkpoint. If the two digests are equal, then the state transferred is correct. E_i then changes the value of *updated* to be equal to *stable*.
2. E_i updates its state to reflect the execution of requests with sequence numbers from *updated* + 1 to $s - 1$. To perform the update, E_i retrieves those requests from the `agree` messages stored in the local *requests* queue, and then actually executes those requests.

Computation of checkpoint digests and state transfer can be made efficient through the use of incremental checkpointing techniques described in [3].

Once a reply certificate has been obtained, it is easy to pinpoint which of the previously implicated replicas (if any) are actually corrupt. A correct replica E_i adds such replicas to its local *corrupt* set, and updates its *pc* accordingly. E_i also shares this information with other replicas, by sending a `convict` message to all replicas. The `convict` message has the form (`convict`, k, s, *proof*), where *proof* contains the reply certificate and replica E_k's `reply` message for `agree(s)`. Once a correct replica has added E_k to its *corrupt* set, it discards any further protocol messages received directly from E_k.

4.4   Primary Committee Changes

At a correct execution cluster replica, any *pc* change is the result of a change in the sets *slow* or *corrupt* and is always accompanied by the sending of `indict` or `convict` messages respectively. Thus, it is not possible for corrupt replicas to force a change in the *pc* when the *pc* replicas are indeed correctly functioning. Those messages contain sufficient proof to convince any other correct execution cluster replica to effect the same change in its own local *slow* or *corrupt* sets. As a result, even though correct replicas may temporarily differ in their perspectives of the primary committee, their perspectives will eventually concur.

4.5   Failure Detection and Its Effect on Protocol Operation

What the parsimony-based audit mode and the recovery mode accomplish when the *pc* is not able to produce a reply certificate is the distributed identification of *pc* replicas that are not functioning properly. The identification is essentially a form of failure detection and is done with the goal of eventually making the *pc* consist only of correct replicas.

In our formal system model, the adversary controls the scheduling of messages and hence the timeouts; thus, the adversary can cause a correctly functioning *pc* replica to be added to the *slow* sets of correct replicas.

Unlike the adversary in our formal model, the network in a real-world setting will not always behave in the worst possible manner. The motivation for an optimistic protocol such as ours is the hope that timer values that are set based on stable network conditions have a high likelihood of being accurate. Such a hope is not unrealistic since practical observations point to the fact that network behavior alternates between long periods of stable conditions and relatively short periods of stability; this indicates that unstable network conditions are the exception rather than the norm. During periods of stability and when the *pc* replicas do not actively misbehave, the optimistic hope will be satisfied and our protocol will provide resource-efficient request execution with roughly the same latency as the all-active approach.

Even if the optimistic hope is not satisfied, our protocol guarantees safety and liveness. Safety mainly relates to replica state consistency. Since replicas always execute a request bound to sequence number s only after a state update that reflects the execution of all lower-sequence-numbered requests, safety is never violated. Liveness, which is the ability to obtain a reply certificate eventually, is also guaranteed; inaccurate failure detection can, at worst, cause correct *pc* replicas to be added to the *slow* sets at correct replicas, but then the protocol will switch to the recovery mode, which guarantees that a reply certificate will be obtained.

Neutralizing the Effect of Inaccurate Failure Detection. Since the adversary corrupts at most t replicas and the only replicas added to the *corrupt* set are those that actually exhibited malicious failures, the *corrupt* set at a correct replica never exceeds t. However, due to inaccurate failure detection, it is possible that correct replicas will get added to the *slow* set, and subsequently, $|slow \cup corrupt|$ may exceed t. To allow the next *pc* to be chosen, whenever $|slow \cup corrupt| = t + 1$, the *slow* set is reset to the empty set, $\emptyset$. A reset counter c is used to keep track of the number of resets. Both **suspect** and **indict** messages carry an indication of the reset counter value. This allows the garbage collection of all `indict` and `suspect` messages with lower reset-counter values, whenever c is incremented.

Since a correct replica E_i sends an `indict` message for each new entry to its local *slow* set and a `convict` message for each new entry to its local *corrupt* set, if E_i encounters a situation in which $|slow \cup corrupt| > t$, then any correct replica E_j will also eventually encounter a situation $|slow \cup corrupt| > t$. Thus, if the reset-counter c at replica E_i is incremented, then eventually all correct replicas will also increment their respective reset-counters to $c + 1$.

5   Protocol Properties

Any replication protocol is required to guarantee safety and liveness. Safety is specified by the total order, update integrity, and result integrity properties described below. Liveness is specified by the termination property described below. Parsimony characterizes the resource efficiency obtained under perceived stable conditions, and distinguishes our protocol from the *all-active* approach. Due to space constraints, we omit the proofs here; they can be found in the full version of the paper [10].

Termination: If the $\mathcal{AC}$ sends `agree`(s), it eventually receives a reply certificate.

Total Order: At any two correct execution cluster replicas E_i and E_j, the updates to their internal states due to execution of the request indicated by `agree`(s) are the same.

Update Integrity: Any correct execution cluster replica updates its internal state in response to the request indicated by `agree`(s) at most once, and only if $\mathcal{AC}$ actually sent that message.

Result Integrity: If r is the result value in the reply certificate received by $\mathcal{AC}$ for `agree`(s), then at least one correct execution cluster replica sent a `reply` message with result value r.

Parsimony: A correct execution cluster replica E_i that is not part of the primary committee will execute the request indicated by the `agree`(s) message and then send a corresponding `reply` message to other replicas only if (1) E_i has not yet obtained a reply certificate for the request, and (2) E_i added s to its local *mustDo* set due to a corrupt or slow replica.

6   Experimental Evaluation

We implemented and experimentally evaluated the parsimony-based execution protocol under both fault-free conditions and controlled fault injections. We compare the results for our protocol with those obtained for the all-active execution approach. All implementations were done in C++.

The fact that the execution phase of a BFT-replicated service will be service-specific poses a challenge to obtaining useful results. The resources involved during request processing will be service-specific, and even request-specific. In our experiments, we have tried to account for that fact by varying the range of service-specific parameters, like the resource intensity of request processing. The specific resource type that we emphasized in our experiments is the CPU, but the conclusions we draw are also an indicator of the trends for other resource types (e.g., network bandwidth) that may be involved in request processing. Our intention was to give a flavor of how parsimony-based execution compares with all-active execution for different service types.

We conducted our experiments for execution cluster sizes $n_e = 3, 5, 7, 9$, and 11 that can tolerate $t = 1, 2, 3, 4$, and 5 simultaneous replica faults, respectively. In a real-world setting, the $\mathcal{AC}$ would consist of a set of $3t+1$ agreement replicas; however, to keep the focus on the execution phase of BFT replication, the clients and the agreement cluster replicas were represented by a single $\mathcal{AC}$ process that generated requests and provided the properties given in Sect. 3.

The setup consisted of a testbed of 12 otherwise unloaded machines. Each machine had a single Athlon XP 2400 processor and 512 MB RAM running RedHat Linux 7.2. One machine was devoted to running the $\mathcal{AC}$ process. At most one execution replica ran on the other machines. The computers were connected by a lightly loaded full-duplex 100 Mbps switched Ethernet network. Digital signatures and MACs were generated using 1024-bit RSA and SHA-1 respectively. Each replica maintained about 1 MB of service-specific state, organized into 1 KB blocks and loaded into its main memory at initialization time.

The $\mathcal{AC}$ sends two kinds of requests: *retrieve-compute* requests and *update-compute* requests. Additionally, for the parsimony-based execution protocol, every 200th $\mathcal{AC}$ request is a checkpoint request. A *retrieve-compute* request specifies a block to be retrieved. A replica performs some computation on the contents of the block, and returns the result in a `reply` message; there is no change to the replica state. An *update-compute* request specifies a block and new contents for the block. A replica updates the specified block with the new contents, performs some computation on the new contents, and returns the result. The argument field of a *retrieve-compute* request is only a few bytes specifying the block number; for an *update-compute* request, the argument field has the size of a block (1 KB). The result field of the `reply` message for either type of request contains the result of the computation and has the size of a block (1 KB). The $\mathcal{AC}$ sends a new request after obtaining a reply certificate for its last request.

6.1   Behavior in Fault-Free Runs

We conducted two sets of experiments that were differentiated by the amount of computation involved in request processing. For the first set of experiments, processing a request involved computation of a public key signature on a specified block of the service state twice; we call such requests *computation-level 2* or *CL-2* requests. For the second set of experiments, processing a request involved computation of a public key signature on a specified block of the service state 100 times; we call such requests *CL-100* requests. Obviously, one would be hard-pressed to find a real-world application that computes digital signatures 100 times for a request. The intention was to simulate compute-intensive request processing (e.g., an insurance web service that has to solve multi-parameter insurance models to obtain results for auto insurance quotation requests), in which the cost of computing one digital signature (in the audit mode of our protocol) is an insignificant part of the actual request processing overhead.

We measured request latency, which is the time elapsed from when the $\mathcal{AC}$ sends a request until it obtains a reply certificate for the request. Figure 3(a) compares the request latencies of the parsimony-based and the all-active execution approaches for CL-2 requests. Figure 3(b) does the same for CL-100 requests. The latencies were obtained as the average of the last 5,000 values from 20 separate runs, where a run consisted of the $\mathcal{AC}$ sending about 10,000 requests. The $\mathcal{AC}$ generated *retrieve-compute* and *update-compute* requests alternately. The latency for a checkpointing request in parsimony-based execution was amortized among all the requests in the corresponding checkpointing interval.

Figures 3(a) and (b) show only a small difference in the request latencies between all-active and parsimony-based execution. For CL-2 requests (Fig. 3(a)), the request latencies for all-active execution are slightly higher than those for parsimony-based execution. The reason is that in all-active execution, even though the $\mathcal{AC}$ needs only $t + 1$ `reply` messages with identical result values to accept the result, it will receive `reply` messages from all replicas (i.e., $2t + 1$ messages), since the runs were fault-free. Though the $\mathcal{AC}$ fully processes only $t + 1$ of those messages and discards the other t, there is overhead involved in receiving the additional unnecessary messages and examining their headers. Thus, one can expect higher latencies for all-active execution if the `reply` message sizes are increased. For compute-intensive CL-100 requests (Fig. 3(b)), the latencies for all-active execution are slightly lower than those for parsimony-based execution. The reason is that all-active execution allows the $\mathcal{AC}$ to choose the fastest $t + 1$ replies among the $2t + 1$ replies that will eventually be received at the $\mathcal{AC}$.

Since in our experiments the CPU is the dominant resource used at a replica in processing $\mathcal{AC}$ requests, we used the UNIX 'ps -aux' command to measure the percentage of CPU utilization on a replica's host machine that is due to request processing. The CPU utilization percentage at a replica was obtained as the average of samplings made every 5 seconds in each run (a run spanned the time it took to process 10,000 $\mathcal{AC}$ requests). The CPU utilization percentages for *pc* replicas in parsimony-based execution and those for any replicas in all-active execution were roughly the same (in the 75%-85% range for CL-2 requests and in the 85%-95% range for CL-100 requests). The CPU utilization percentages for backups in parsimony-based execution were negligible for both CL-2 and CL-100 requests.

After obtaining the average CPU utilization percentages at the individual replicas, we computed the *overall CPU utilization factor*, which we obtained by summing over all replicas the product of the CPU utilization percentage

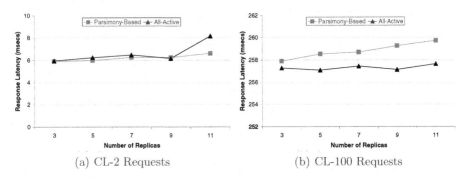

(a) CL-2 Requests   (b) CL-100 Requests

Fig. 3. Request Latency

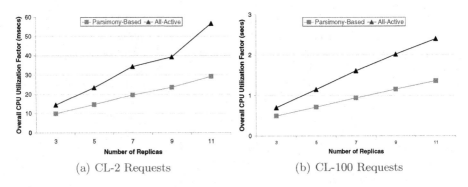

(a) CL-2 Requests   (b) CL-100 Requests

Fig. 4. Overall CPU Utilization Factor Per Request

and the time taken to process a request. Figures 4(a) and (b) show the overall CPU utilization factor for CL-2 and CL-100 requests. The utilization factor for parsimony-based execution is roughly half of that for all-active execution, and the reduction is more pronounced as the number of replicas increases. This is a practically significant result. For example, in the Application Service Provider (ASP) business model (see Section 7.1), the overall CPU utilization factor would be an indicator of the total amount of CPU resources spent by the ASP servers per request, and could form the basis for pricing, especially if request processing is compute-intensive.

6.2   Behavior in the Presence of Fault Injections

We conducted fault injection experiments on our protocol. We did not fault-inject the implementation of all-active execution, since its behavior in the presence of faults would not be much different from its behavior when there are no faults.

Figure 5(a) shows the different factors that contributed to the $\mathcal{AC}$ request latency when a pc member was fault-injected after servicing a sufficiently large number of CL-2 requests (about 5,000). We injected both *muteness* faults and *malicious* faults in our protocol. A muteness fault injection was done by crashing

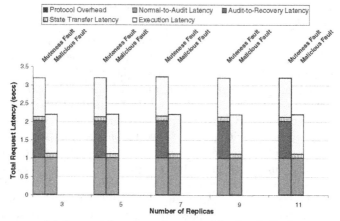

(a) Request Latency when a *pc* Member is Faulty

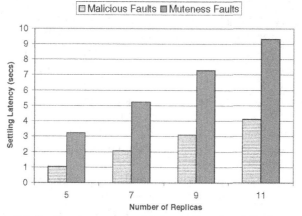

(b) Settling Latency for Multiple Correlated Faults

Fig. 5. Behavior Under Fault Injections

a *pc* member upon receipt of a specified $\mathcal{AC}$ request. A malicious fault injection was done by making a *pc* member send wrong values in its `reply` messages.

The highest latency is obtained when a muteness fault injection is done. The latency comprises two timeout values, state update latency, and the protocol-specific overhead. The first timeout value of 1 second (represented by "normal-to-audit latency" in the graph) is used at the $\mathcal{AC}$ before the $\mathcal{AC}$ sends a *retransmit* message for the request. Receipt of that message will cause the protocol to switch from parsimony-based normal mode to parsimony-based audit mode. The second timeout value of 1 second (represented by "audit-to-recovery latency" in the graph) causes a replica to send a `suspect` message for the crashed *pc* member, as the replica would not have received a `reply` message from the *pc* member for the request. Once $n - t$ `suspect` messages for the crashed *pc* member have been received, the protocol switches from the parsimony-based audit mode to recovery mode. In the recovery mode, backups bring their states up to date in two steps before sending their own `reply` messages for the $\mathcal{AC}$ request. The first step (represented by "state transfer latency" in the graph) is the transfer of state from a correct pc member up to the last stable checkpoint. The second step (represented by "execution latency" in the graph) is the actual execution of all the requests after the checkpoint request up to the request for which $\mathcal{AC}$ sent a *retransmit* message. To bring out the worst-case behavior, we injected the muteness fault into a pc member upon receiving the request just prior to the checkpoint request (so that $\delta - 1$ requests would actually have to be executed), and all backups requested state transfer from the same correct pc member. The portion marked "protocol overhead" in the graph includes a round-trip transmission time from the $\mathcal{AC}$ to the replica (for the request message from the $\mathcal{AC}$ and the `reply` message from the replica) plus other overhead related to the parsimony-based protocol (such as exchange of `suspect` and indict messages, and selection of a new pc). We see that an overwhelmingly large portion of the request latency when a muteness fault is injected depends on tunable system parameters (like timeout) and service-specific values, such as the size of the application state, the checkpointing technique used, the number of requests beyond the last stable checkpoint that have to be executed to bring the state up to date, and the normal request processing latency. The actual overhead due to the parsimony-based protocol is less than 20 milliseconds.

The $\mathcal{AC}$ request latencies for malicious fault injection are essentially the $\mathcal{AC}$ request latencies for muteness fault injection minus the timeout value used at replicas (i.e., the audit-to-recovery latency). Fault detection is much faster for malicious faults because it is based on examination of the contents of the `reply` message rather than on timeouts.

Figure 5(b) quantifies the effect that multiple correlated fault injections have at the replicas. After servicing a sufficiently large number of requests (about 5,000), we injected multiple faults at the replicas so that a new pc member fault was activated every time an $\mathcal{AC}$ request arrived until the fault resiliency t of the replication group was exhausted. As before, the $\mathcal{AC}$ sent a request only after accepting a result for its previous request. We injected both muteness and malicious faults, and thus there are two rows of bars in the graph. The first fault was activated at a checkpoint request to bring out the worst-case behavior. At each correct replica, we measured the *settling latency*, i.e., the time from when the first fault is detected at a replica until the time when the pc consists only of non-fault-injected replicas. The time includes the fault detection latency for $t-1$ faults (i.e., for all faults except the first fault), the state update latency, the time to execute t $\mathcal{AC}$ requests, and the overhead due to the parsimony-based execution protocol. Since the multiple faults are activated at consecutive requests, backups have to bring their state up to date only once, after the first fault detection.

As expected, both rows of bars in the graph show an increase in the settling latency as the number of replicas (and hence the number of fault injections, t) increases. For a given t, the settling latency for muteness faults is higher than that for malicious faults. The reason is that the fault detection latency for t

muteness fault injections includes $2(t - 1)$ timeouts (the factor of 2 being due to the $\mathcal{AC}$ timeout plus the timeout at replicas), as opposed to only $(t - 1)$ $\mathcal{AC}$ timeouts for t malicious fault injections.

7   Discussion

In this section, we give examples of applications that would benefit from our protocol and compare our protocol with related work.

7.1   Practical Applications

Our protocol can yield significant benefits in many applications. Below are two examples:

1. The web service infrastructures for many companies are no longer operated by the companies themselves, but are outsourced to third parties called *Application Service Providers* or *ASPs*. The ASPs own, operate, and maintain the servers running the applications that provide the companies' web services, saving the companies the cost burden of having to set up specialized information technology infrastructures. The ASPs' servers may be shared among several companies. Usually, an ASP charges an outsourcing company a consumption fee based on the actual resource use. In such a situation, BFT replication can be very useful in enhancing the trustworthiness of computations, and our protocol can be used to obtain significant reductions in overall execution costs and thereby the fee that the outsourcing company has to pay to the ASP. The benefits are especially pronounced if the web service application is resource-intensive. An example of a web service for which request processing is computation-intensive would be a financial web service that has to solve multi-parameter financial models to predict stock trends.
2. In the computational Grid, many services are computation-intensive. BFT replication can be used to obtain a trustworthy system from untrusted participating Grid nodes. Since Grid nodes may be shared among several Grid services, our protocol can help significantly reduce the performance impact that one service has on other services running in the same Grid node.

7.2   Related Work

BFT replication techniques are of two categories: quorum replication and state machine replication. Quorum replication (e.g., [11]) uses subsets of replicas (called *quorums*) to implement read/write operations on the variables of a data repository, such that any two subsets intersect in enough correct replicas. State machine replication can be used to perform arbitrary computations accessing arbitrary numbers of variables; quorum replication is less generic and cannot handle concurrent requests by clients to update the same information. Our protocol is similar to quorum systems in that it uses a subset of replicas to perform operations. However, the similarity is only superficial; our protocol is concerned with the execution phase of state machine replication, our use of a $(t+1)$-subset of replicas to execute requests is based on whether the system is stable or not (a distinction that quorum systems do not make), and (unlike quorum systems) we do not use different subset sizes for read and write operations.

Our protocol is both unique and novel. While most work on BFT replication has focused on the hard problem of Byzantine agreement (e.g., [3,12]), our work focuses on the often-overlooked but practically significant execution phase of BFT replication. Yin et al.'s work reduces the *deployment costs* of BFT replication by reducing the number of execution replicas from $3t+1$ to $2t+1$. However, our work deals with reducing the *run-time or operational costs* of BFT replication, which are likely to be at least as important as deployment costs in many long-lived and resource-intensive applications. While the parsimony principle has been routinely used in primary-backup systems that tolerate benign faults (e.g., [13]), our protocol is novel in that it is the first to apply parsimony to Byzantine fault tolerance.

Since our protocol is for the execution cluster, our work is complementary to the BASE work [4] and the BASE extension by Yin et al. [5]. In particular, one could combine our protocol with (1) the proactive recovery and abstraction techniques of BASE to overcome the drawbacks of state machine replication in many applications (namely, the determinism requirement and the assumption that at most one-third of the replicas are corrupt), and (2) the privacy firewall architecture of [5] to obtain BFT confidentiality.

In the context of parallel computing, Sarmenta [14] proposed mechanisms for tolerating erroneous results submitted by malicious volunteers in the Grid, SETI@home, and other *volunteer computer systems*. The mechanisms, called *credibility-based fault-tolerance* mechanisms, estimate the credibility of a node and use these probability estimates in limiting the amount of redundant computations necessary to meet desired error rates. However, their scheme trades off correctness for performance and is not relevant to applications that are stateful or cannot tolerate any errors at all (e.g., banking or financial applications). Also, their mechanisms operate in a system and fault model that is very restrictive (e.g., it requires synchronous computations and non-collusion among malicious nodes) and less generic than our system and fault model.

8   Conclusion

We described a protocol for executing requests in a resource-efficient way while providing trustworthy results in the presence of up to t Byzantine faults. Previous best solutions were based on the all-active approach, which requires at least $2t+1$ replicas to execute a request. Our protocol reduces service-specific resource use costs to about half of what they are for all-active execution under perceived normal conditions by using only a *pc* consisting of $t+1$ execution replicas to execute the request. The benefits are more pronounced for larger group sizes, and when request processing is resource-intensive. The trade-off for the benefits is the higher latencies during perceived failure or instability conditions due to fault detection and service-specific state update latencies. It is reasonable to expect that a system's operation will alternate between long periods of normal conditions and short periods of instability. That motivated our decision to optimize our protocol for the common case, even if it means paying a slightly higher cost during periods of instability.

Acknowledgments. We thank Christian Cachin, Kaustubh Joshi, and Ryan Lefever for many insightful discussions and valuable suggestions for improving the quality of the paper. We thank Jenny Applequist for her editorial comments.

References

1. Schneider, F.B., ed.: Trust in Cyberspace. National Academy Press (1999)
2. Lamport, L.: Time, Clocks and Ordering of Events in Distributed Systems. Communications of the ACM **21** (1978) 558–565
3. Castro, M., Liskov, B.: Practical Byzantine Fault Tolerance and Proactive Recovery. ACM Transactions on Computer Systems (TOCS) **20** (2002) 398–461
4. Rodrigues, R., Castro, M., Liskov, B.: BASE: Using Abstraction to Improve Fault Tolerance. In: Proceedings of the 18th Symposium on Operating System Principles. (2001) 15–28
5. Yin, J., Martin, J.P., Venkataramani, A., Alvisi, L., Dahlin, M.: Separating Agreement from Execution for Byzantine Fault Tolerant Services. In: Proc. 19th Symp. on Operating Systems Principles. (2003) 253–267
6. Kursawe, K.: Optimistic Byzantine Agreement. In: Proc. 21st Symposium on Reliable Distributed Systems. (2002) 262–267
7. Vanstone, S.A., van Oorschot, P.C., Menezes, A.: Handbook of Applied Cryptography. CRC Press (1996)
8. Cachin, C., Kursawe, K., Petzold, F., Shoup, V.: Secure and Efficient Asynchronous Broadcast Protocols. In: Advances in Cryptology: CRYPTO 2001 (J. Kilian, ed.), LNCS-2139, Springer (2001) 524–541
9. Goldwasser, S., Micali, S., Rivest, R.L.: A Digital Signature Scheme Secure Against Adaptive Chosen-Message Attacks. SIAM Journal on Computing **17** (1988) 281–308
10. Ramasamy, H.V., Agbaria, A., Sanders, W.H.: A Parsimonious Approach for Obtaining Resource-Efficient and Trustworthy Execution. Submitted for publication in the IEEE Transactions on Dependable and Secure Computing (2005)
11. Malkhi, D., Reiter, M.: Byzantine Quorum Systems. Journal of Distributed Computing **11** (1998) 203–213
12. Reiter, M.K.: The Rampart Toolkit for Building High-Integrity Services. In: Selected Papers from the International Workshop on Theory and Practice in Distributed Systems, LNCS 938, Springer-Verlag (1995) 99–110
13. Budhiraja, N., Schneider, F., Toueg, S., Marzullo, K.: The Primary-Backup Approach. In Mullender, S., ed.: Distributed Systems, ACM Press - Addison Wesley (1993) 199–216
14. Sarmenta, L.F.G.: Sabotage-tolerance mechanisms for volunteer computing systems. Future Generation Computer Systems **18** (2002) 561–572

Generating Fast Atomic Commit from Hyperfast Consensus*

Fabíola Gonçalves Pereira Greve[1] and Jean-Pierre Le Narzul[2,3]

[1] Computer Science Department, Federal University of Bahia, Brazil
fabiola@ufba.br
[2] GET/ENST-Bretagne, 35512 Cesson-Sévigné, France
JP.LeNarzul@enst-bretagne.fr
[3] IRISA, Campus de Beaulieu, 35042 Rennes, France

Abstract. This work introduces a highly modular derivation of fast
non-blocking atomic commit protocols. Modularity is achieved by the
use of consensus protocols as completely independent services. Fast decision is obtained by the use of consensus protocols that decide in one
communication step in good scenarios. Two original non-blocking atomic
commit protocols are presented. One of the presented protocols outperforms existing equivalent solutions that are based on the use of failure
detectors. In the presence of a low resiliency rate, $f \leq 1$, it behaves as
the classical 2PC and 3PC, exhibiting the same message complexities.
In the general case, when one considers the number of tolerated crashes
as $f < n/2$, it exhibits a complexity of $2nf + 3n$ point to point messages.
The best known algorithm exhibits a complexity of $4nf + 3n$ point to
point messages.

1 Introduction

The concept of *transaction* is used in distributed systems or databases to ensure consistent actions on distributed data. An *atomic commit* (AC) protocol is
at the heart of a transactional system; such a protocol guarantees, in the presence of failures, the *failure atomicity* (also called *all-or-nothing*) property of the
transaction: either every process *commits* or every process *aborts*. Of course, the
outcome of a transaction depends on the local conditions at every process' site.
When a process can locally make permanent the modifications to data (e.g. no
concurrency control conflict has been detected), it reveals its intention by voting
YES. The outcome of a transaction depends on the collected votes. If all processes
vote YES, the outcome will be COMMIT. Otherwise, the outcome will be ABORT.

The well-known two-phase commit protocol (2PC) [J78] is the simplest
atomic commit protocol and the one that exhibits the best performance: *three*
communication steps and $3n$ point to point messages are enough to commit.
Unfortunately, in presence of faults, it blocks. A non-blocking protocol allows

* This work is supported by CNPQ/Brazil and by the cooperation project
CAPES/COFECUB 497/05.

C.A. Maziero et al. (Eds.): LADC 2005, LNCS 3747, pp. 226–244, 2005.

correct processes to take decisions even in the occurrence of faults. *Three-Phase Commit* protocols (3PC) [S81, KD95] are non-blocking, but, besides strong synchrony requirements, they exhibit a high latency to finish: *five* communication steps and $5n$ point to point messages.

The non-blocking *atomic commit* problem (NB-AC) belongs to the class of *agreement* problems where processes belonging to a same group have, from time to time, to reach an unanimous decision. A weak version of it, namely, nonblocking weak atomic commit (NB-WAC), can be solvable in an asynchronous model augmented with unreliable failure detectors [CT96, Gue95]. In this case, it can be reduced to a more basic problem, known as the consensus problem. The consensus is defined in terms of two primitives: *propose* and *decide*. Each process proposes an initial value and then executes a consensus algorithm till a unique value is decided. Protocols solving the atomic commit problem have been proposed from reductions to the consensus [GS95, R97, HT97, GLS96, GL04]. They are non-blocking and exhibit a lower latency than the 3PC protocols. Unfortunately, in these protocols, modularity competes with efficiency. The modular solutions require an important number of broadcasted messages to tolerate failures: $O(n^2)$ [GS95, R97, HT97]. The best protocols exhibit a message complexity of $O(nf)$ messages [GLS96, GL04], if one considers f as the maximum number of processes that may crash. To reach these good performances, they make use of *ad-hoc* protocols, where the consensus and the atomic commit are wrapped in an unique block in which protocols are inseparable.

The first contribution of this paper is the proposal of an elegant approach to design modular and efficient non-blocking atomic commit protocols. In our approach, an atomic commit protocol relies on a hyperfast consensus protocol that decides in one communication step in good scenarios. When good scenarios do not apply, the hyperfast consensus protocol makes use of an underlying consensus that allow it to terminate. In this general schema, consensus is used as a termination protocol for the atomic commit protocol only when necessary, in case of failures or erroneous suspicions. The main advantage of this approach is that, when certain good but realistic conditions are satisfied, an efficient solution to the consensus problem directly leads to fast atomic commit protocols. By *fast*, we mean algorithms that decide a transaction (abort or commit) as soon as possible. This happens when some process votes NO (resulting in abort) or all processes vote YES and there are no failures or erroneous suspicions (resulting in commit).

From the proposed schema, we have derived two atomic commit protocols, namely AC-Set and AC-Value. AC-Set (respectively AC-Value) relies on a hyperfast consensus protocol, called Set-Consensus (respectively Val-Consensus). These two consensus have been proposed by one of the authors in [BGMR01, Gre02]. They introduce new consensus assumptions giving rise to hyperfast algorithms, that allow the learning of decided values within one communication step. This happens when a sufficient number of processes propose the same value for consensus. Recently [Lam04], Lamport has pointed out the importance in studying and applying new pertinent consensus definitions as a way for breaking the limit on message delays for agreement problems. This article contributes to investigate this approach by the derivation of high-performance atomic commit protocols. Besides, it shows that the design of well structured protocols is compatible with high performance.

Thus, the second contribution of this paper is the proposal of atomic commit protocols that exhibit performances that equal or overcome those of ad-hoc protocols proposed so far. Both proposed protocols are as efficient as the (2PC) in terms of latency. They terminate after *three* communication steps in the absence of failures. Protocol AC-Set is more efficient than any other failure-detector based AC protocol published so far. It requires $(2nf + 3n)$ messages (without a broadcast network) or $(n + f + 2)$ messages (with a broadcast network) to commit. Thus, in presence of a low resiliency rate $(f < 2)$, it is as efficient as 2PC and 3PC. The best known protocol, proposed by Guerraoui *et al.* [GLS96] requires $(4nf + 3n)$ messages (without a broadcast network) or $(n + 2f + 2)$ messages (with a broadcast network). Protocol AC-Value requires $(n(2n + 1))$ messages (without a broadcast network) or $(2n + 1)$ messages (with a broadcast network). The protocol proposed by Guerraoui *et al.* in [GS95] exhibits the same complexity as AC-Value but it requires stronger conditions to early decide. In the case of AC-Value, decisions are sped up if at least $(f + 1)$ propositions are for COMMIT in the consensus phase. In the case of [GS95], decisions are taken only if *every* correct process proposes COMMIT.

Let us remark that very recently, Dutta, Guerraoui and Pochon in [DGP04] investigate the time-complexity of the NB-AC problem in a synchronous environment. They propose ad-hoc protocols that either fast abort or fast commit a transaction when no processes crash. Their algorithms are close to our AC-Value protocol, but for a synchronous model. Our algorithms are built for the asynchronous model.

The rest of this paper is organized as follows. In Section 2, we define the system model and the consensus problem. In Section 3, we present the non-blocking atomic commit problem. Section 4 introduces two consensus algorithms that can decide in one communication step: Val-Consensus and Set-Consensus In Section 5, we describe how we use one-step consensus algorithms to derive two efficient solutions to the atomic commit problem: AC-Value and AC-Set. In Section 6, we analyze the cost of our AC protocols. Section 7 compares the performance of our AC protocols with others in terms of latency and number of messages. Finally, Section 8 concludes this paper. The appendix contains the correctness proof of the AC-Set protocol.

2   Distributed System Model

Asynchronous System. The system model is patterned after the one described in [CT96, FLP85]. It consists of a finite set Π of $n > 1$ processes, namely, $\Pi = \{p_1, \ldots, p_n\}$. A process can fail by *crashing*, i.e., by prematurely halting; a crashed process does not recover. A process behaves correctly (i.e., according to its specification) until it (possibly) crashes. By definition, a *correct* process is a process that does not crash. A *faulty* process is a process that is not correct. As indicated in the introduction, f denotes the maximum number of processes that may crash. Processes communicate and synchronize by broadcasting and receiving messages through channels. Communication is reliable: there is no message corruption, duplication or loss. If a process crashes while broadcasting a message m, only a subset of processes can receive m. There are assumptions neither on the relative speed of processes nor on message transfer delays. *One communication step* is characterized by the emission and the corresponding reception of a set of messages.

The Consensus Problem. In the traditional consensus problem, every process p_i *proposes* a value v_i and all correct processes *decide* on some unique value v, in relation to the set of proposed values. More precisely, it is defined by the following properties [CT96, FLP85]:

- C-Termination: every correct process eventually decides some value;
- C-Validity: if a process decides v, then v was proposed by some process;
- C-Uniform-Agreement: no two processes (correct or not) decide differently [1].

Unfortunately, the consensus problem is actually impossible to solve in a deterministic way in asynchronous distributed systems when even a single process may crash. This is known as the Fischer-Lynch-Paterson (FLP) impossibility result [FLP85].

Unreliable Failure Detectors. To circumvent the consensus impossibility result, several approaches have been investigated. One of them is based on unreliable failure detectors proposed by Chandra and Toueg [CT96]. Failure detectors can be classified according to the properties (completeness and accuracy) they satisfy. A class of failure detectors denoted $\Diamond S$ is of particular interest because it has been proved to be the weakest one that allows to solve the consensus problem [CHT96]. This class is defined by the following completeness and accuracy properties: Any process that crashes is eventually suspected (*strong completeness*) and there is a time after which there is a correct process that is no longer suspected (*eventual weak accuracy*). Relying on $\Diamond S$ failure detectors to solve agreement problems assumes that a majority of processes within the system never fails. Note that in an asynchronous model extended with unreliable failure detectors, whenever consensus is solvable, uniform consensus is equally solvable [Gue95].

Reliable Broadcast. A reliable broadcast [HT93] primitive ensures the atomic delivery of a message by every correct process. Informally, a reliable broadcast of message m (implemented by function R-Broadcast (m)) guarantees that m is delivered (implemented by function R-Deliver (m)) by all correct processes if the sender of m is correct or by all correct processes or none of them if the sender of m is not correct.

---

[1] We consider here the uniform version of the consensus problem in which agreement is reached even for faulty processes. In the classical consensus problem, this property holds only for the correct processes.

3   The Non-blocking Atomic Commit Problem

In Section 1, we have informally defined the NB-AC problem. Now, we give the formal properties specifying it:

- (i) AC_Termination: every correct process eventually decides;
- (ii) AC_Uniform_Agreement: no two processes decide differently;
- (iii) AC_Validity: the decision is COMMIT or ABORT. Besides, to avoid trivial decisions where processes decide independently of the collected votes, one states:

  - AC_Justification: if a process decides COMMIT then every process has voted YES;
  - AC_Obligation: if every process votes YES and *there is no failure*, the decision is COMMIT.

3.1   Solving the NB-AC Problem in Asychronous Systems

The two-phase commit protocol (2PC) [J78] is the simplest AC protocol and the one that exhibits the best performance (three communication steps are enough to commit). The coordinator requests votes from the processes participating to the transaction. If every process votes YES, the coordinator broadcasts a COMMIT decision. Otherwise, i.e. one process votes NO or one process is suspected to be faulty, the coordinator broadcasts an ABORT decision. Unfortunately, in the presence of failures, 2PC is blocking (i.e. the property AC_Termination is violated); this is due to the fact that it is based on a centralized coordinator. The failure of the coordinator may prevent non-failed processes, waiting for a decision from the coordinator, to decide the outcome of the transaction. In such a situation, non-failed processes are blocked and cannot release resources they previously acquired.

Consequently, it is desirable to derive non-blocking AC protocols, that are able to take decisions in the presence of failures. The three-phase commit protocols [S81, KD95] (3PC) are such protocols. However, they are not without serious drawbacks. Their first drawback is the cost (higher latency than 2PC): they need five communication steps and the broadcast of $5n$ messages. The second one is the complexity of the quorum (majority of processes) based recovery procedure used to terminate when the coordinator fails [R97]. Moreover, these protocols require reliable failure detectors.

Guerraoui [Gue95] studied the problem of NB-AC in the context of an asynchronous model extended with unreliable failure detectors. He showed that the NB-AC problem is more difficult to solve than the consensus. The consensus problem is solvable in an asynchronous model extended with unreliable failure detectors whereas the NB-AC problem is not. This result comes from the AC_Obligation property that requires to reliably detect failures (which is impossible with unreliable failure detectors). A solution is to replace the AC_Obligation condition by a weaker condition that leads to the definition of a weaker problem, called NB-WAC (Non-Blocking Weak Atomic Commit).

3.2   The Non-blocking Weak Atomic Commit Problem

The NB-WAC problem keeps the same definitions for all the properties of the NB-AC problem but the AC_Obligation: if all the processes vote YES and *no process is ever suspected*, then the decision is COMMIT. One interesting characteristic about the NB-WAC problem is its reductibility to the consensus problem [Gue95]. Thus, the results obtained for solving the consensus in an asynchronous model with unreliable failure detectors apply for solving the NB-WAC problem. Several protocols solving the atomic commit problem were obtained from reductions to the consensus [GS95, R97, HT97, GLS96, GL04]. In Section 6 we compare some of these protocols with the solutions suggested in this paper.

4   Hyperfast Consensus

Theoretical results showed that one cannot solve the classical consensus problem in less than two communication steps [CBS00, KR01]. An algorithm that achieves this bound is known as *early deciding*. In a recent publication [Lam04], Lamport has pointed out the interest in studying new pertinent consensus definitions as a way for breaking the limit on message delays. In [BGMR01, Gre02], one of the authors proposed new consensus assumptions giving rise to hyperfast consensus algorithms, that allow the learning of decided values within only one communication step. Those assumptions basically consist in enriching the initial knowledge of processes with an *a priori* agreement, besides the pair (n, f). The practical interest of these protocols is demonstrated in this paper through the derivation of very efficient atomic commit protocols. We recall in this section, these new consensus families that allow one step decision when $f < n/2$.

Underlying Principle of Hyperfast Consensus. The idea that underlies the design of our protocols is very simple. It comes from the following observation: if all the processes initially propose the same value, then this value is necessarily the decided value, whatever the protocol and the system behavior. Hence, the suggested protocols execute a first communication step during which the processes exchange the values they propose. Afterwards, each process checks whether some of the processes have the same initial value. If it is the case, this value is decided. If it is not, an underlying consensus protocol is used.

Underlying Consensus Protocol. Our aim is to provide a consensus protocol that terminates in one communication step in good scenarios but also terminates in bad scenarios. So, we consider that the underlying asynchronous distributed system allows to solve the consensus problem. More precisely, we assume it is equipped with a black box solving the consensus problem, and we provide a protocol that decides in one communication step in good scenarios and uses the underlying consensus protocol in the other cases. A process p_i locally invokes it by calling Underlying_Consensus(v_i) where v_i is the value it proposes.

4.1   Consensus Guided by a Privileged Value

Condition. Let α be a distinguished value in the set of values that can be proposed. Moreover, let us assume that α is initially known by each process. The *a priori* knowledge of such a value can help expedite the decision when $f < n/2$ as shown in Fig. 1. The idea of the protocol is simple: a process is allowed to decide α in one communication step as soon as it knows that α has been proposed by at least $f + 1$ processes.

Function Val_Consensus(v_i, α)

Task $T1$:
% ——————— Phase 1: Early Deciding
(1)   *send* PROPOSED(v_i) to all $p_j : p_j \in \Pi$;
(2)   **wait until** (PROPOSED messages received from $(n - f)$ processes)
                    or (received $f + 1$ values equal to α);
(3)   **if** $(f + 1$ of the received values are equal to $\alpha)$ **then**
(4)   R-Broadcast (C-DECISION(α));
% ——————— Phase 2: Deciding by Underlying Consensus
(5)   **else**
(6)   **if** $(\alpha$ received from a process) **then** $v_i \leftarrow \alpha$ **endif**;
(7) $return($Underlying_Consensus$(v_i))$;
(8)   **endif**

Task $T2$:
(9)   **upon** R-Deliver of C-DECISION(v) **do** $return(v)$;

Fig. 1. Consensus Guided by a Privileged Value (Val_Consensus)

Behavior. A process p_i begins execution by calling Val_Consensus(v_i, α). The function ends when it carries out the *return* command with the decided value (lines 7 or 9). As usual, in order to prevent the blocking of a process (waiting for a value from another process that has already decided), a process that decides, uses a reliable broadcast to disseminate its decision value. So, the function Val_Consensus() is made up of two tasks: $T1$ and $T2$. $T1$ implements the core of the protocol. Line 4 and $T2$ make use of the reliable broadcast functions. Task $T1$ begins by a first phase, where processes broadcast and collect their proposals (lines 1-2). When a process p_i learns that α has been proposed by at least $(f + 1)$ processes, then p_i can safely decide α (lines 3-4). To decide at this phase, processes do not have to call upon a failure detector service.

Processes that do not decide in one step can adopt α as their proposal value in a second phase (line 6). This is possible because, being given that $(f + 1)$ proposals with α exist, then, any process necessarily receives at least 1 PROPOSED message carrying out the α value. So, whenever $(f + 1)$ process proposes the same value α, *all* the processes which do not decide in line 4, call upon Underlying_Consensus in line 7 with the same value α. Therefore, necessarily, α is the decided value for everyone.

4.2   Consensus Guided by a Set of Participants

Condition. Let us consider the existence of a set $S \subset \Pi$, whose composition is known a priori by every process. In other words, there is a group of processes which are not anonymous for the computation. If all the processes belonging to S propose the same value, then it is possible to decide in one communication step when $f < n/2$. The protocol described in Fig. 2 uses this strategy and requires $|S| > f$. In practice, the set S should be chosen in order to gather the most reliable servers of the system or the fastest ones, since they will be responsible for the early deciding.

Function Set_Consensus(v_i, S)

Task $T1$:
% ——————— Phase 1: Early Deciding
(1) **if**   ($p_i \in S$) **then** *send* PROPOSED(v_i) to all $p_j : p_j \in \Pi$; **endif**
(2) **wait until**
        (($\forall\, p_j \in S : p_j \in suspected\_i$ **or** PROPOSED message received from p_j))
        and ($\exists\, p_j \in S$: PROPOSED message received from p_j))
(3) **if** (the same value v has been received from each process $\in S$) **then**
(4)   R-Broadcast (C-DECISION(v));
% ——————— Phase 2: Deciding by Underlying Consensus
(5) **else** $v_i \leftarrow$ a value from a process $\in S$;
(6)   *return*(Underlying_Consensus(v_i));
(7) **endif**

Task $T2$:
(8) **upon** R-Deliver of C-DECISION(v) **do** *return*(v);

Fig. 2. Consensus Guided by a Set of Participants (Set_Consensus)

Behavior. The protocol is shown in Fig. 2. It behaves as Val_Consensus() (Fig. 1). However, only the processes belonging to S take part in the broadcast of the suggested value (line 1). Then, all the processes await for values coming from S (line 2). In order to unblock the protocol, we called upon the failure detector service. Since $|S| > f$, at least one value from a member of S will be received by all processes. Whenever p_i certifies that a same value v was proposed by *all* the processes belonging to S, then it can safely decide v in a single communication step (lines 3-4). As the previous protocol, the safety properties are assured because, since there is at least $(f + 1)$ processes in S ($|S| > f$), any process necessarily receives at least (1) PROPOSED message coming from a process of S. So, these processes can adopt v as their proposition value in line 5. Therefore, when all processes of S propose the same value v, *all* the processes that do not decide in line 4, invoke Underlying_Consensus in line 6 with the same value v. Necessarily, v is the settled value.

5   Fast Atomic Commit Protocols

In this section, we describe the design of a generic and modular solution to the atomic commit problem based on the hyperfast consensus algorithms presented in Section 4. We have derived two efficient AC algorithms from this generic solution: (1) atomic commit guided by a value, (2) atomic commit guided by a set of participants. The functions involved in the implementation of the atomic commit protocols are shown in Fig. 3. The Transaction() function relies on an Atomic_Commit() function that relies itself on a Hyperfast_Consensus() function. Each of these modules accesses a list $suspected_i$ given by a failure detector associated to the process.

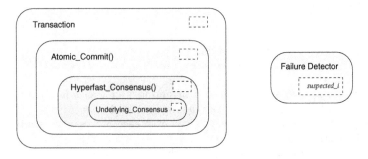

Fig. 3. Hierarchy of the Functions Involved in the Atomic Commit Protocols

Transaction's Module. This module (see Fig. 4) implements a transaction. It is run by every process. One of those processes, known as the *leader*, is in charge of coordinating the decision procedure for the transaction. The leader initiates the protocol by asking every process p_i to declare its intention to validate operations on data (lines 1-2). A process p_i sets its vote v_i to *yes* (line 6) if it is locally ready to make the updates permanent; it sets its vote to *no* if it is not locally ready (line 7) or if it suspects the leader (line 8). Then, every process p_i returns the result of the call to Atomic_Commit($vote_i$) (line 9).

This function implements a NB-AC protocol and ensures a unique result for the transaction.

Atomic Commit's Module. The Atomic_Commit() protocol uses the consensus service as a black box. It is made of two phases. During the first phase, every process broadcasts its vote and collects the votes from the other processes (votes *yes* or *no*). Depending on the collected votes, every process starts a second phase by running a hyperfast consensus algorithm to early decide. They broadcast a COMMIT or ABORT proposition for the transaction. The algorithm can terminate immediately after this second phase if some "good conditions" are met. These conditions are specific to each consensus protocol and depend on the collected votes. Processes that have not been able to terminate start a third phase and run an underlying consensus to decide a uniform result.

Procedure Transaction
(1) **if** $(p_i = leader)$ **then**
(2)   send REQUEST_VOTE<> to all $p_j : p_j \in \Pi$; **endif**
(3) **wait until** (REQUEST_VOTE<> received from *leader*
            or *leader* $\in suspected_i$);
(4) **if** (REQUEST_VOTE<> received from *leader*) **then**
(5)   **if** (able to make updates permanent)
(6)   **then** $vote_i \leftarrow$ YES;
(7)   **else** $vote_i \leftarrow$ NO; **endif**
(8) **else** $vote_i \leftarrow$ NO; **endif**
(9) *return* Atomic_Commit($vote_i$);

Fig. 4. Transaction's Module

By using the principle given above and the early decision ability of the consensus algorithms, we have derived two solutions to the NB-WAC problem. The first reduction to Val_Consensus(), called AC-Value(), is shown in Fig. 5. The privileged value α being selected to COMMIT, it decides as soon as a sufficient number of processes choose COMMIT. A second reduction to Set_Consensus(), called AC-Set(), is shown in Fig. 6. A subset S of processes is selected in advance; the decisions are taken as soon as all the processes from S propose the same value for the outcome of the transaction.

Behavior. During the first phase, every process broadcasts its vote to others. If a process p_i cannot locally commit to make the updates permanent ($vote_i = no$), it reliably broadcasts its decision and immediately decides to abort the transaction (line 2) in a unilateral way. This decision is legitimate because ABORT is the only acceptable decision with respect to the properties of the problem. A reliable broadcast is necessary to ensure the agreement and the termination of the computation (Line 2 and the concurrent task $T2$ make use of the reliable broadcast functions). A process that is ready to validate the transaction ($vote_i = yes$), after a first phase of vote exchanges, starts a second phase with one of the consensus algorithms presented in Section 4. These algorithms early terminate in the first phase (that corresponds to the second phase for the atomic commit protocol).

5.1 Atomic Commit Guided by a Value

The function AC-Value (Fig. 5) has an intuitive behavior. During the *first phase*, *every* process broadcasts its vote and collects the votes from the others (lines 1 and 3). During the *second phase*, every process starts Val_Consensus(v, α) and proposes a value v that depends on the collected votes. If a process was able to collect positive votes from all the others, it proposes $v =$COMMIT (lines 4-5), otherwise it proposes $v =$ABORT (line 6) as the first parameter to the consensus.

In most of the runs, processes are able to validate the transaction and succeed in gathering the positive votes from the others. So, in such runs, COMMIT is the

---

Function Atomic_Commit($vote_i$)

Task T1:

% ——————— Phase 1: Exchange of Votes

(1) *send* VOTE< $vote_i$ > to all $p_j : p_j \in \Pi$;

(2) **if** ($vote_i$ = NO) **then** R_Broadcast AC-DECISION <ABORT>; **endif**

(3) **wait until** ($vote_j$ received from all $p_j \in (\Pi - suspected_i)$);

% ——————— Phase 2: Exchange of propositions

(4) **if** ($\forall\, p_j \in : \Pi$ ($vote_j$ received) **and** ($vote_j$ = YES))

(5) **then** *return* Val_Consensus (COMMIT, *commit*);

(6) **else** *return* Val_Consensus (ABORT, *commit*); **endif**

Task T2:

(7) **upon** R-Deliver of AC-DECISION < v > **do** *return*(v);

Fig. 5. Atomic Commit by Value (AC-Value)

value proposed by every process. We can take benefit from this observation to ensure early termination and thus consider *commit* as the privileged value α for the consensus. So, if a sufficient number of processes ($f + 1$) vote COMMIT, it is possible to decide at the first phase of Val_Consensus() (lines 1-4, Fig. 1). The processes that do not decide, start in a *third phase*, the Underlying_Consensus algorithm in order to obtain a uniform result for the transaction.

5.2   Atomic Commit Guided by a Set of Participants

A participant that is ready to locally commit ($vote_i = yes$) initiates the protocol only with the members of set S. So, during the *first phase*, every participant

---

Function Atomic_Commit($vote_i$)

Task T1:

% ——————— Phase 1: Exchange of Votes

(1) *send* VOTE< $vote_i$ > to all $p_j : p_j \in S$;   % $|S| > f$, $S \subseteq \Pi$ %

(2) **if** ($vote_i$ = NO) **then** R_Broadcast AC-DECISION <ABORT>; **endif**

(3) **if** ($p_i \in S$) **wait until** ($vote_j$ received from all $p_j \in (\Pi - suspected_i)$);

% ——————— Phase 2: Exchange of propositions

(4) **if** ($p_i \in S$) **then**

(5)   **if** ($\forall\, p_j \in \Pi$: ($vote_j$ received) **and** ($vote_j$ = YES))

(6)   **then** *return* Set_Consensus (COMMIT, S);

(7)   **else** *return* Set_Consensus (ABORT, S);   **endif**

(8) **else** *return* Set_Consensus ($\bot$, S); **endif**

Task T2:

(9) **upon** R-Deliver of AC-DECISION < v > **do** *return*(v);

Fig. 6. Atomic Commit by a Set (AC-Set)

sends its vote to the members of S (line 1); members of S wait for the votes from the non-suspected (line 3). During the *second phase*, depending on the collected votes, processes belonging to S start the consensus with COMMIT or ABORT (lines 5-7). The other processes $(\Pi - S)$ do not participate to the initial proposal for Set_Consensus(v_i) (line 1, Fig. 2). So, they call the function with a non significant value $(v = \perp)$ (line 8). If every process from S proposes the same value (COMMIT or ABORT), then it is possible to decide during the first phase of Set_Consensus() (lines 1-5, Fig. 2). In Section 6, we show that the selective broadcast of votes to the members of S leads to good performances for our protocol. The processes that do not decide, start, in a *third phase*, the Underlying_Consensus algorithm.

6   Cost of Atomic Commit Protocols

We present the cost of our protocols in a favorable scenario: there is neither failures, nor erroneous suspicions and all processes validate the transaction (vote *yes*). It is the most frequent case in practice. As we explained previously (Section 3), in the presence of crashes, our solutions enjoy the same advantages associated with the use of the consensus as a termination protocol. We measure the number of communication steps and the number of necessary sent messages to decide. We are interested in the exchange of messages in two different environments: i) point to point network and ii) broadcast network. In our model, a message broadcasted to all in environment (i) has cost n^2. In environment (ii) it has cost 1.

Fig. 7 illustrates the phases and the number of communication steps achieved by the atomic commit protocols[3]. Both of them decide in *three communication steps*. The first step is necessary to start the transaction (execution of the transaction service). Then, every module Atomic_Commit() finishes in two communication steps: one step to distribute the votes (*yes* or *no*) and another step to distribute the propositions (COMMIT or ABORT). The computation ends in the following conditions: (i) protocol AC-Value() requires that $(f + 1)$ propositions are equal to COMMIT; (ii) protocol AC-Set() requires that *all* processes in S adopt the same proposition (they have identical values, either for COMMIT, or for ABORT). These conditions are perfectly achieved when the favorable scenario described in the previous paragraph occurs.

Protocol AC-Value (Fig. 7 (a)), exhibits a number of point to point messages equal to $n(2n + 1)$ (or $2n + 1$ in a broadcast environment). This result is obtained by the sum of the following values:

---

[2] The actual complexity is $n - 1$ instead of n. But, for the sake of clarity, we are not considering this absolute value.

[3] When taking into account the complexity of these protocols, we do not consider the message complexity due to the use of the *reliable broadcast* primitive[CT96] which is inherent to consensus protocols. This is the current practice in the literature since in a fault free scenario, asymmetric protocols do not rely on the reception of this message to decide.

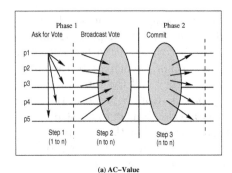
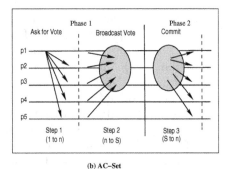

(a) AC–Value   (b) AC–Set

Fig. 7. Communication Steps of Protocols with $leader = p_1$ and $S = \{p_1, p_2, p_3\}$

- first step: n point to point messages (or 1 broadcast message), that represents the demand of votes by the transaction leader to all;
- second step: n^2 point to point messages (or n broadcast messages), that represents the exchange of votes between everybody;
- third step: n^2 point to point messages (or n broadcast messages), that represents the first phase of the consensus in which all the processes exchange their propositions.

Recall that the cardinality of S is $> f$, therefore for the sake of efficiency, one can consider $|S| = f + 1$. Thus, protocol AC-Set (Fig. 7 (b)) exhibits a number of point to point messages equal to $2nf + 3n$ (or $n + f + 2$ in a broadcast environment). This number is the sum of the following values:

- first step: n point to point messages (or 1 broadcast message), that represents the demand of votes by the transaction leader to all;
- second step: $n(f + 1)$ point to point messages (or n broadcast messages), that represents the sending of votes from everybody to processes belonging to S;
- third step: $n(f + 1)$ point to point messages (or $f + 1$ broadcast messages), that represents the first phase of the consensus in which processes in S send their propositions to everybody.

7   Related Work and Comparison

Several works considering the use of the consensus service as a terminating protocol to deal with crashes and solve this problem have been suggested elsewhere [GS95, R97, HT97, GLS96, GL04]. In the following, we compare the AC-Value and AC-Set protocols with some of these works. Table 1 summaries the complexities exhibited by the classical 2PC, 3PC and by consensus-based protocols that decide in three communication steps. We consider the same favorable scenario and the same communication environment as the one described in Section 6.

Table 1. Performance Measurements of Atomic Commit Protocols

Class	Protocol	# Comm. Steps	# Point-to-Point Msgs	# Broadcast Msgs
Classical	2PC	3	$3n$	$n+2$
	3PC	5	$5n$	$2n+3$
Failure Detector-	[GS95]	3	$n(2n+1)$	$2n+1$
	AC-Value	3	$n(2n+1)$	$2n+1$
Based	[GLS96]	3	$4nf+3n$	$n+2f+2$
	AC-Set	3	$2nf+3n$	$n+f+2$
Leader-Based	FasterPaxosCommit	3	$2nf+3n$	$n+f+2$

Latency and Message Complexity. The 2PC [S81] requires only *three* communication steps and the emission of $3n$ point to point messages to decide. However, it does not solve the NB-WAC problem. The non-blocking 3PC [KD95] protocols accomplish the decision with two extra communication steps but require the sending of $5n$ point to point messages. Protocols presented by Hurfin and Tronel [HT97] and Raynal [R97] terminate in at least four communication steps and require $O(n^2)$ messages to commit.

The protocol proposed by Guerraoui and Schiper [GS95] presents the same performance results as AC-Value. In favorable conditions, they early decide in *three* communication steps and they require the sending of $n(2n+1)$ point to point messages. To the best of our knowledge, the work of Guerraoui *et al.* [GLS96] is the failure-detector-based protocol which presents the best performance in a good scenario. It finishes in *three* communication steps and requires the sending of $4nf+3n$ point to point messages (or $n+2f+2$ broadcast messages). Our protocol AC-Set outperforms [GLS96] results. It has the same latency but requires the diffusion of $2nf+3n$ point to point messages (or $n+f+2$ broadcast messages). So, in relation with [GLS96], it saves $2nf$ point to point messages[4].

Recently, Gray and Lamport have published Paxos-based commit protocols [GL04, Lam98, Lam01] exhibiting good performance results. One algorithm, called *Paxos Commit* requires one more message delay than 2PC, so it decides in *four* communication steps. On the other hand, it requires to send less messages than AC-Set: only $nf+3n+f$ messages are necessary to commit. Another algorithm, called *Faster Paxos Commit* [GL04], is an optimized version to reduce message latency. It exhibits the same message complexity as AC-Set.

Similarities and Differences. The recent commit protocols proposed by Gray and Lamport [GL04] are based on a Paxos consensus [Lam98, Lam01]. The Paxos consensus assumes some method of choosing a *leader*. So, differently from fail-

---

[4] The definition of message complexity adopted by Guerraoui *et al.* in [GLS96] takes into consideration the necessary number of *received* messages to reach commit. Differently, we count the number of *sent* messages. We think that in this way we reproduce more faithfully the number of messages that effectively transit in the net. Therefore, in their paper, instead of showing the complexity of $4nf+3n$ *sent* messages, [GLS96] shows a complexity of $3nf+3n$ *received* messages to decide.

ure detectors consensus, liveness properties are satisfied if the leader-selection algorithm ensures that a unique non faulty leader is chosen whenever a set of enough processes are non faulty during a sufficient period. AC-Set is based on a one-step consensus algorithm which ensures termination if a sufficient number of processes propose the same value for the agreement. In both solutions, progress is assured if at least $f + 1$ coordinating processes are active. So, even if these protocols are based on different consensus approaches, they need the same conditions to converge.

In spite of the similarities of protocols [GS95] and AC-Value, we can verify that the first one requires strong conditions to early decide. In the case of AC-Value, decisions are speed up if at least $(f + 1)$ propositions are for COMMIT. In the case of [GS95], decisions are taken only if *every* correct process proposes COMMIT. Protocols [GLS96] and AC-Set adopt the same strategy: the early termination is guided by a special set of participants. In [GLS96], a set of processes, named Set_{NB} is chosen *a priori* and is responsible for leading the protocol to an early decision. Set_{NB} plays the same role as the set S in protocol AC-Set. Protocols can be distinguished in the way the set is used and in the way the decision is accomplished. In both solutions, the messages from the first phase (exchange of votes *yes* or *no*) are only broadcasted to the processes belonging to the set (S or Set_{NB}). Afterwards, in a second phase, only processes in the set broadcast the propositions to validate the transaction (COMMIT or ABORT). At this point, in both solutions, the processes can decide if they gather the good conditions for this. If they do not early decide, they start a third phase, calling upon a consensus service. In the case of AC-Set, *every* process can participate in the consensus. In protocol [GLS96], only the processes belonging to Set_{NB} will participate. Thus, to ensure liveness, $|Set_{NB}| \geq 2f+1$. This distinction carries the secret of the best performance results of protocol AC-Set, which requires $S \geq f + 1$. Interestingly, this lower bound coincides with the lower number of coordinators necessary to make *Faster Paxos Commit* progress. It uses multiple coordinators and make progress if a majority of them are alive. So, $2f + 1$ coordinators are required and they can make progress even if f of them are faulty.

Protocol [GLS96] early decides only if processes belonging to Set_{NB} vote for COMMIT. AC-Set allows an early decision in a more general situation: if processes in S vote equally, either for COMMIT or for ABORT. Let us observe that this is very desirable in presence of erroneous suspicions or process faults after the first phase (after the exchange of votes). That means, all processes are for validating (vote *yes*) at the time of the first phase, but thereafter, due to suspicions or faults, they propose ABORT to the consensus. Even in the presence of such a scenario our solution anticipates the decision.

Highly Modular Derivations. The design strategy used to derive our AC protocols is highly modular. The consensus is a completely independent service that should be regulated by specific parameters, in order to guarantee *hyperfast* decisions in favorable conditions. Additionally, if the early decision phase does not succeed, any other underlying consensus can be used to continue the computation. This modular structure is not observed in most AC protocols because, generally, modularity competes with efficiency. This is the case for [GLS96] and *Faster Paxos Commit*. Their design integrate the consensus and the atomic commit in an unique block in which protocols are inseparable. Moreover, the *framework* obtained from the proposed resolution of the atomic commit is generic enough to be used in the solution of other agreement problems (such as, atomic broadcast [CT96], group membership [GHRT01], etc.). That is currently being investigated by the authors.

General Evaluation. In general, consensus-based protocols exhibit point to point message complexities of $kn^2 + \mathrm{G}(n)$ and broadcast messages of $k'n$. The best know protocols ([GLS96], AC-Set, *Paxos Commit* and *Faster Paxos Commit* have results that are dependent on the number f of tolerated faults. In the case $f < n/2$, it is interesting to note that AC-Set requires $n^2 + \mathrm{G}(n)$ point to point messages and $3/2n$ broadcast messages, whereas the others, except *Paxos Commit*, require $2n^2 + \mathrm{G}(n)$ and $2n$. Besides, in the presence of a low resiliency rate, AC-Set is as efficient as 2PC (in a failure free scenario) or 3PC, in presence of one fault.

8 Conclusion

The problem of *atomic commit* is essential in the implementation of distributed transactions, since it is in charge of guaranteeing the data consistency in spite of the occurrence of faults. In this work, we presented original atomic commit protocols based on new fast deciding consensus algorithms. One of the obtained protocols is more efficient than any other failure detector-based protocol proposed so far [GLS96].

In the absence of faults, it exhibits the same behavior as 2PC and in the presence of 1 fault, it exhibits the same message complexity as 3PC. We succeeded in defining both efficient and modular protocols. Modularity is obtained thanks to the organization of the consensus as a completely independent block of the atomic commit that uses its service. Moreover, we think that the framework designed for our solutions is generic enough to be adapted to other specializations in order to solve other agreement problems in which the proposed values are similar.

Acknowledgments

The authors would like to thank Michel Hurfin from IRISA Labs at Rennes (France) for its valuable contribution to this work.

References

[BGMR01]   F. Brasileiro, F. Greve, A. Mostefaoui and M. Raynal, Consensus in One Communication Step is Possible. In *PaCT (International Conference on Parallel Computing Technologies)*, Springer-Verlag LNCS 1800, pp. 1258-1265, September 2001.

[CBS00]   B. Charon-Bost and A. Schiper, *Uniform Consensus is Harder than Consensus*. Technical Report, École Polytechnique Fédérale de Lausane, Switzerland, DSC/2000/028. May, 2000.

[CHT96]   T. Chandra, V. Hadzilacos and S. Toueg, The Weakest Failure Detector for Solving Consensus. *Journal of the ACM*, 43(4):685–722, July 1996.

[CT96]   T. Chandra and S. Toueg, Unreliable Failure Detectors for Reliable Distributed Systems. *Journal of the ACM*, 43(2):225-267, March 1996.

[DGP04]   P. Dutta, R. Guerraoui and B. Pochon, *Fast non-blocking atomic commit: an inherent trade-off*. Inf. Process. Lett. Vol. 91, no 4, pages 195–200. 2004.

[FLP85]   M. Fischer, N. Lynch and M. Paterson, Impossibility of Distributed Consensus with One Faulty Process. *Journal of the ACM*, 32(2):374–382, April 1985.

[GHRT01]   F. Greve, M. Hurfin, M. Raynal, F. Tronel, Primary Component Asynchronous Group Membership as an Instance of a Generic Agreement Framework. *ISADS'2001: 5th International Symposium on Autonomous Decentralized Systems*, pp 93-100, March 2001.

[GL04]   J.Gray and L. Lamport, *Consensus on Transaction Commit*, Technical Report, Microsoft Corporation, MSR-TR-2003-96, January, 2004.

[GLS96]   R. Guerraoui, M. Larrea and A. Schiper, Reducing the Cost for Non-Blocking in Atomic Commitment. In *16th International Conference on Distributed Computing Systems*, pp. 692-697, Hong-Kong, May 1996.

[Gre02]   F. Greve, *Réponses efficaces au besoin d'accord dans un groupe*, PhD Thesis, Univ. of Rennes I. Nov. 2002.

[GS01]   R. Guerraoui and A. Schiper, The Generic Consensus Service. *IEEE Transactions on Software Engineering*, 27(1), pp. 29-41, January 2001.

[GS95]   R. Guerraoui and A. Schiper, The Decentralized Non-Blocking Atomic Commitment Protocol. In *14th IEEE International Symposium on Parallel and Distributed Processing*, pp. 2-9, San Antonio, October 1995.

[Gue95]   R. Guerraoui, Revisiting the Relationship Between Non-Blocking Atomic Commitment and Consensus. In *9th Int. Workshop on Distributed Algorithms*, Springer-Verlag, LNCS 972, pp. 87–100, (J-M. Hélary and M. Raynal Eds), Le Mont-Saint-Michel (France), September 1995.

[HT93]   V. Hadzilacos and S. Toueg, In *Distributed Systems, ch Fault Tolerant Broadcasts ans Related Problems*. pp. 97-145, 1993

[HT97]   M. Hurfin and F. Tronel, A Solution to Atomic Commitment Based on an Extended Consensus Protocol. In *6th IEEE Workshop on Future Trends of Distributed Computing Systems*, pp. 98-103, 1997.

[J78]   J. Gray, Notes on Database Operating Systems. In *Operating Systems An Advanced Course*, pp 10-17. Lecture Notes in Computer Science (60), Springer-Verlag. 1978.

[KD95]   I. Keidar and D. Dolev, Increasing the Resilience of Atomic Commit, at No Additional Cost. In *ACM PODS'1995: Principles of Database Systems*, pp 245-254, May 1995.

[KR01]   I. Keidar and S. Rajsbaum, *On the Cost of Fault-Tolerant Consensus when There are No Faults: a Tutorial*, MIT Technical Report, MIT-LCS-TR-821, 2001. Preliminary version in *SIGACT News, Distributed Computing Column* (2001), 32(2):45-63.

[Lam98]   L. Lamport, The part-time parliament, In *ACM Transactions on Computer Systems*, pp 133-169, 16(2), May 1998

[Lam01]   L. Lamport, Paxos Made Simple, *ACM SIGACT News,Distributed Computing Column*, 32(4):18-25. Dec. 2001.

[Lam04]   L. Lamport, *Lower Bounds for Asynchronous Consensus*, Technical Report, Microsoft Corporation, MSR-TR-2004-72, July 2004.

[R97]   M. Raynal, Revisiting the Non-Blocking Atomic Commitment Problem in Distributed Systems. In *2nd Workshop on Fault-Tolerant Parallel and Distributed Systems*, 1997.

[S81]   D. Skeen, NonBlocking Commit Protocols. In *ACM SIGMOD International Conference on Management of Data*, pp 133- 142. ACM Press. 1981.

A   Correctness Proof

We sketch here a brief correctness proof of AC-Set (see Fig. 6). Correctness proof of AC-Value is similar and thus it will not be described here. In the following, we are assuming the correction of Set_Consensus [BGMR01, Gre02].

Theorem 1. AC_Validity: *the decision is* COMMIT *or* ABORT.

Proof. In function Atomic_Commit processes decide for ABORT in task $T2$ (line 9) and $T1$ (line 2). All the other decisions are a consequence of the execution of the Set_Consensus in lines 6, 7 and 8. The C_Validity of the consensus guarantees that a decided value is a proposed one. From line 1 of Set_Consensus, only processes belonging to S propose a value. Thus, from lines 6 and 7 of Atomic_Commit, only COMMIT or ABORT will be proposed. This proves the Theorem. $\square_{Theorem\ 1}$

Theorem 2. AC_Justification: *the decision is* COMMIT *only if all participants voted* yes.

Proof. In task $T2$ (line 9) and $T1$ (line 2), the function Atomic_Commit only decides for ABORT. Other decisions are a consequence of the execution of the Set_Consensus in lines 6, 7 and 8. The value COMMIT is proposed to the consensus only in line 6. This happens if all of the processes of Π vote *yes* (line 5). The C_Validity of the consensus guarantees that a decided value is a proposed one. This proves the Theorem. $\square_{Theorem\ 2}$

Theorem 3. AC_Obligation: *if every process votes* yes *and there is no* failure supicion *then the decision is* COMMIT.

Proof. To accomplish the proof, we will show that the value ABORT can only be decided if some process votes *no* or if it suspects the failure of another process. In function Atomic_Commit, a participant decides ABORT in lines 2, 9 and in the execution of the Set_Consensus function (lines 6-8). In this case,

- p_i decides in line 2 only if it votes *no*;
- if p_i decides in line 9, another process p_j has executed line 2. Thus, p_j has voted *no*;
- when ones executes function Set_Consensus and decides for the ABORT value, somebody must have proposed ABORT; this comes from the C_Validity property of consensus which states that a decided value must have to be a proposed value. In this case, a process must have executed line 7 proposingABORT. From lines 3-5, this process must have suspect some other process. This proofs the Theorem. $\square_{Theorem\ 3}$

Theorem 4. AC_Uniform_Agreement: *No two participants decide differently.*

Proof.

- In the Atomic_Commit function, a process decides ABORT in task $T2$ (line 9) only and only if another process p_j voted *no* in task $T1$. In consequence, p_j has decided for ABORT in the task $T1$ (line 2).
- If some process decided ABORT in line 2, from lines 5-7, those processes that continue the execution invoke the Set_Consensus proposing ABORT (if they are in S) (line 7) or $\perp$ (those not belonging to S). Once only propositions from processes in S are considered in the execution of Set_Consensus (lines 1, 2 and 6), the processes that decide in the consensus unit, will decide for ABORT.
- If no process decides in line 2, all the correct processes continue the execution and invoke Set_Consensus. From the C_Uniform_Agreement property of consensus, two processes do not decide differently. So, the Theorem follows. $\qquad \square_{Theorem\ 4}$

Theorem 5. AC_Termination: *all of the correct processes decide in a definitive way.*

Proof. The only instructions that could block a correct process to keep the protocol's execution are:

- Lines 3 (from the transaction unit Transaction) and 3 (from the Atomic_Commit function). However, from the *strong completeness* property of the failure detector $\Diamond S$, every faulty process will be eventually suspected. Moreover, every process begins the Atomic_Commit function by broadcasting its vote(line 1). Since channels are reliable, every vote from a correct process will eventually be received. Thus, the correct processes do not block in the execution of these instructions.
- Calling the Set_Consensus (line 6- 8) service. From property C_Termination of consensus, every correct process definitely decides. Thus the Theorem follows. $\qquad \square_{Theorem\ 5}$

Group-Based Replication of On-Line Transaction Processing Servers*

A. Correia Jr., A. Sousa, L. Soares, J. Pereira, F. Moura, and R. Oliveira

Computer Science and Technology Center,
Computer Science Department, University of Minho

Abstract. Several techniques for database replication using group communication have recently been proposed, namely, the Database State Machine, Postgres-R, and the NODO protocol. Although all rely on a totally ordered multicast for consistency, they differ substantially on how multicast is used. This results in different performance trade-offs which are hard to compare as each protocol is presented using a different load scenario and evaluation method.

In this paper we evaluate the suitability of such protocols for replication of On-Line Transaction Processing (OLTP) applications in clusters of servers and over wide area networks. This is achieved by implementing them using a common infra-structure and by using a standard workload. The results allows us to select the best protocol regarding performance and scalability in a demanding but realistic usage scenario.

1 Introduction

Synchronous database replication provides both transparent distribution and fault tolerance. By keeping data strictly up-to-date in all replicas, application programmers do not have to manage complex reconciliation procedures and fail-over can happen without causing any committed updates to be lost. Recently, replication techniques based on group communication have been proposed as a means to overcome performance bottlenecks and make synchronous replication cost-effective [1,12,11,17,15,20].

In contrast with replication based on distributed locking and atomic commit protocols, group communication based protocols minimize interaction between replicas and the resulting synchronization overhead by relying on total order multicast to ensure consistency. Generically, the approach builds on the classical replicated state machine [19]: The exact same sequence of update operations is applied to the same initial state, thus producing a consistent replicated output and final state. The problem is then to ensure deterministic processing without overly restricting concurrent execution, which would dramatically reduce throughput, and avoid re-execution in all replicas.

These concerns have been addressed by several proposals based on group communication [14,15,17,13]. Although all rely on a totally ordered multicast for consistency, they differ mainly in whether transactions are executed conservatively [14,15] or optimistically [17,13]. In the former, by a priori coordination among the replicas, it is assured that when a transaction executes there is no concurrent conflicting transaction

---

* Supported by FCT, project STRONGREP (POSI/CHS/41285/2001).

C.A. Maziero et al. (Eds.): LADC 2005, LNCS 3747, pp. 245–260, 2005.

being executed remotely and therefore its success depends entirely on the local database engine. In the latter ones, execution is optimistic, each replica independently executes its locally submitted transactions and only then, just before committing, sites coordinate and check for conflicts between concurrent transactions.

This difference results in multiple and often subtle performance and resiliency trade-offs. Namely, how does each protocol cope with a large share of update transactions, conflicting updates, high latency in wide area networks, and symmetric load to multiple replicas. Unfortunately, each protocol is presented using a different load scenario and evaluation method which makes it very hard to clearly highlight the main consequences of the approach.

In this paper we evaluate the suitability of group based replication protocols for replication of On-Line Transaction Processing (OLTP) applications in clusters of servers and over wide area networks (WAN). This evaluation compares the protocols using a common infrastructure which rests on a novel common database interface suitable for the implementation of group based replication protocols. Using the same settings for all protocols and the workload of the industry standard TPC-C benchmark [23] it is possible to establish relative strengths and select the best protocol for each scenario.

The rest of the paper is structured as follows. In Sect. 2, we briefly review the main group-based database replication approaches. Section 3 introduces the common implementation and evaluation framework. Section 4 presents and discusses performance measurements. Finally, Sect. 5 concludes the paper.

2   Replication Protocols

In this study, we consider three replication protocols: CONS, a protocol that implements the conservative execution approach (similar to those proposed in [14,15]) and two protocols that exploit optimistic execution, Postgres-R (PGR) [13] and the Database State Machine (DBSM) [17]. All of these protocols are multi-master, transactions can be submitted to and executed by several replicas, and follow the passive replication paradigm [4,8], each transaction is executed by one of the replicas and its state changes propagated to the other replicas.

At the core of these protocols is a total order (or atomic) multicast primitive [10]. Some of the proposed algorithms [16,14,15,20] have been presented using total order primitives with optimistic delivery. The goal is to compensate the inherent ordering latency by allowing tentative processing in parallel with the ordering protocol. If the final order of the messages matches the predicted order then the replication protocol can proceed, otherwise the results obtained tentatively are discarded. We opted, however, not to consider such optimization. In a local area network (LAN), the small message delays discourage any optimistic processing and, in WAN, an algorithm such as the one presented in [21] is required to compensate the large differences and variability of point-to-point latencies. The use of such an algorithm would evenly benefit all of the replication protocols under study but would not contribute to expose the key factors that differentiate them.

The database engine considered implements a multi-version concurrency control mechanism [3]. While locally, the database engine does not provide serializability as its correctness criterion, globally the replication protocols under study are able to do so. In our tests, we will consider and compare both the global 1-copy-serializability[3] and the snapshot-isolation[2] versions of the protocols.

Only update transactions are handled by the replication protocol. Queries are simply executed locally at the database to which they are submitted and do not require any distributed coordination.

In the following, we describe the conservative and optimistic execution approaches and the required interfaces with the database engine.

2.1 Conservative Execution

In the conservative approach, data is a priori partitioned in conflict classes, not necessarily disjoint. Each transaction has an associated set of conflict classes (the data partitions it accesses) which are assumed to be known in advance. While the conflict classes for a transaction could be determined at runtime, this would require to know the whole transaction before its execution precluding the processing of interactive transactions.

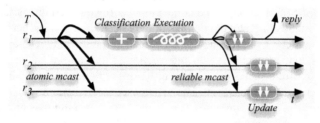

Fig. 1. Conservative replication protocols: CONS

When a transaction is submitted (Figure 1), its id and conflict classes are atomically multicast to all replicas obtaining a total order position. Each replica has a queue associated with each conflict class and, once delivered, a transaction is classified according to its conflict classes and enqueued in all corresponding queues. As soon as a transaction reaches the head of all of its conflict class queues it is executed. Transactions are executed by the replica to which they are submitted.[1]

Conflicting transactions are executed sequentially. Clearly, the conflict classes have a direct impact on the performance. The fewer the number of transactions with overlapping conflict classes, the better the interleave among transactions. As we shall discuss in Sect. 2.3, conflict classes are usually defined at the table level but can have a finer grain at the expense of a non-trivial validation process to ensure that a transaction does not access conflict classes that were not previously specified.

When the commit request is received, the outcome of the transaction is reliably multicast to all replicas along with the replica's state changes and a reply is sent to

---

[1] When isolated conflict classes exist, dedicating a distinguished replica to the execution of all transactions of such classes, results in a faster processing of those transactions[15].

the client. Each replica applies the remote transaction's updates with the parallelism allowed by the initially established total order of the transaction.

It is worth noting that, despite the use of a multi-version database engine, since conflicting transactions are totally ordered and executed sequentially, the protocol ensures 1-copy-serializability as long as transactions are classified by the application taking into account read/write conflicts. Relaxing the correctness criterion to snapshot-isolation would simple require the reclassification of the transactions taking into account just write/write conflicts.

2.2   Optimistic Execution

In the optimistic approach, transactions are immediately executed by the replicas to which they are submitted without any a priori coordination. Locally, transactions are synchronized according to the specific concurrency control mechanism of the database engine.

Upon receiving the commit request, a successful transaction is not readily committed. Instead, the tuples read (read-set) and written (write-set) are gathered and a termination protocol initiated. The goal of the termination protocol is to decide the order and the outcome of the transaction such that the global correctness criteria is satisfied. This is achieved by establishing a total order position for the transaction and certifying it (i.e., checking for conflicts) against concurrently executed transactions. The certification of a transaction is done by evaluating the intersection of its read-set and write-set (or just write-set in case of the snapshot-isolation criterion) with the write-sets of concurrent, previously ordered transactions.[2] The fate of a transaction is therefore determined by the termination protocol and a transaction that would locally commit may end up aborted.

The two optimistic protocols, PGR and DBSM (Figure 2), ensure global serializability, but differ in their termination protocols. Both use the transaction's read-sets for the certification procedure. Basically, in PGR the transaction's read-set is not propagated and thus only the replica executing the transaction is able to certify it. In the DBSM, the transaction's read-set is propagated allowing each replica to autonomously certify the transaction.

In detail, upon the reception of the commit request for a transaction t, in PGR the executing replica atomically multicasts t's id and t's write-set and write-values (the values of the tuples in the write-set). As soon as t is ordered, the executing replica certifies t and reliably multicasts the outcome to all replicas. The certification procedure consists in checking t's read-set and write-set against the write-sets of all transactions committed locally since t's commit request.[3] The executing replica then commits or aborts t locally and replies to the client. Upon the reception of the remote transaction's commit outcome each replica applies t's state changes through the execution of a *high priority*

---

[2] The formal definition and detailed explanation of the certification procedures can be found in [13,16,24].

[3] In the original protocol [13], a locking concurrency control mechanism was considered for the database engine which allowed to carry the certification process inside the database as part of the normal execution of the transaction. The read-set was not actually extracted and it consisted of the read locks granted to the transaction.

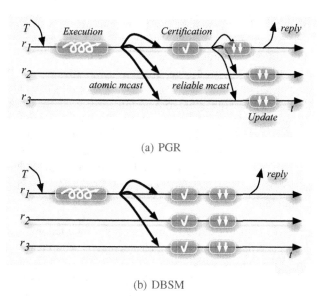

(a) PGR

(b) DBSM

Fig. 2. Optimistic replication protocols

transaction consisting of updates, inserts and deletes according to t's previously multicast write-values. The high priority of the transaction means that it must be assured of acquiring all the required write locks, possibly aborting any locally executing transactions.

The termination protocol in the DBSM is significantly different and works as follows. Upon the reception of the commit request for a transaction t, the executing replica atomically multicasts t's id, the version of the database on which t was executed,[4] and t's read-set, write-set and write-values. As soon as t is ordered, each replica is able to certify t on its own.

For the certification procedure, in the DBSM each replica compares its database version with that of t: if they match t commits, otherwise t's read-set and write-set are checked against the write-sets of all transactions committed locally since t's database version. If they do not intersect, t commits, otherwise t aborts. If t commits then its state changes are applied through the execution of a high priority transaction consisting of updates, inserts and deletes according to t's previously multicast write-values. Again, the high priority of the transaction means that it must be assured of acquiring all the required write locks, possibly aborting any locally executing transactions. The executing replica replies to the client at the end of the transaction.

Of particular relevance for the performance of these two protocols is the definition and representation of the transaction's read-sets. First, read-sets determine the outcome of a transaction certification. If the considered read-set is larger than the set of tuples actually read by the transaction then spurious aborts may arise. On the contrary, if the read-set does not contain the tuples actually read, then serializability may be compro-

---

[4] The database version is a counter maintained by the replication protocol that is incremented every time a transaction commits.

mised. Second, in the DBSM protocol the size of the read set may have a serious impact on the network bandwidth. PGR avoids the propagation of the transaction's read-set at the expense of an additional communication step.

When considering the snapshot-isolation correctness criterion, then both protocols can be simplified and end up being the same. To satisfy snapshot-isolation, certification does not need to check read-write conflicts and thus the transactions' read-sets are not required. As such, the PGR protocol can be simplified by enabling a simpler write-write certification at all the replicas and eliminating the second communication step conveying the outcome of the transaction [24]. The DBSM protocol can also be simplified by not propagating the read-sets and using the simpler certification procedure.

2.3   Database Interface

The replication protocols just described require specific interactions with the adopted database engine. Despite their differences, their interaction with the database engine is similar and the interface can be generalized.

Transactions are submitted to the database engine and evolve through three different phases (Figure 3): the *pre-execute* phase which includes the "begin transaction" command and extends up to the transaction's first statement, the *execute* phase encompassing the whole transaction execution up to the "commit transaction" command, and the *commit* phase from the "commit transaction" command until the reply to the client application. Interactions between the database engine and the replication protocol happen between these three phases and require extended functionality from the database engine.

In the CONS protocol, at the pre-execute phase the database engine needs to be informed about the conflict classes of the transaction. Usually, such classes are defined at the table level to ease the validation process that occurs at the execution phase to ensure that no other classes beyond those specified at the pre-execute phase are accessed. For finer grains, the process would be more complicated. If instead of whole tables, conflict classes were defined using table partitioning such as filters over attributes, guaranteeing that the accessed items are a sub-set of the conflict classes would ultimately lead to a satisfiability problem [9].

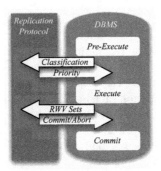

Fig. 3. Interface between the replication protocol and the database engine

In the optimistic protocols, just before entering the execution phase, a remote transaction is assigned high priority allowing it to break any locks currently granted to other transactions. This interaction is required to ensure the successful execution of the updates of remote transactions. Thus, the concurrency control mechanism of the database engine needs to be extended to distinguish these high priority transactions.

After the local execution of the transaction, before the commit phase, the database engine is required to provide the read-set, write-set and write-values (RWV sets). The write-set and write-values are easily extracted from any database engine but the extraction of the read-set requires close coupling with it. While for simple SPJ statements (i.e., statements that involve Select, Project and Join operations) one extraction step is sufficient, more complex queries require the analysis of the optimizers execution plan, multiple extraction points and further read sets combination. Both PGR and the DBSM protocols rely on the transaction's read-set. A judicious characterization and extraction of the read-set is due to avoid unnecessarily large read-sets and consequent spurious aborts, and to reduce network consumption in the case of the DBSM protocol.

Finally, while naturally the outcome of a transaction is decided inside the database engine, with the optimistic replication protocols the fate of a local transaction ultimately depends on the certification procedure. Therefore, it is required that the database engine allow the replication protocol to determine the commit or abort of the transaction.

With respect to the implementation of the necessary functionality of the database engine most of it needs to be done in core. While one could be tempted to implement these interfaces using a middleware approach through the use of triggers, some, such as the pre-execute and commit interfaces, are not possible with current database engines, and others, such as the extraction of the read-sets would lead to unbearable performance hits.

3   Experimental Procedure

This section presents the simulation environment used to evaluate the protocols. We use a centralized simulation model that combines real software components with simulated hardware, software and environment components to model a distributed system. This allows us to setup and run multiple realistic tests with slight variations of configuration parameters that would otherwise be impractical to perform, specially if one considers a large number of replicas and wide-area networks. The centralized nature of the system allows for global observation of distributed computations with minimal intrusion as well as for control and manipulation of the experiment. All tests are conducted under an implementation that mimics the industry standard on-line transaction processing benchmark TPC-C [23].

3.1   Simulation Infrastructure

To evaluate the protocols we use a hybrid simulation environment that combines simulated and real components [22]. The key components, the replication and the group communication protocols, are real implementations while both the database engine and the network are simulated.

In detail, we use a centralized simulation runtime based on the standard Scalable Simulation Framework (SSF) [6], which provides a simple yet effective infrastructure for discrete-event simulation. Simulation models are built as libraries that can be reused. This is the case of the SSFNet [7] framework, which models network components (e.g. network interface cards and links), operating system components (e.g. protocol stacks), and applications (e.g. traffic analyzers). Complex network models can be configured using these components, mimicking existing networks or exploring particularly large or interesting topologies.

To combine the simulated components with the real implementations the execution of the real software components is timed with a profiling timer [18] and the result is used to mark the simulated CPU busy during the corresponding period, thus preventing other jobs, real or simulated, to be attributed simultaneously to the same CPU. The simulated components are configured according to the equipment and scenarios chosen for testing (Sect. 3.2).

The database server handles multiple clients and is modeled as a scheduler and a collection of resources, such as storage and CPUs, and a concurrency control module. The database provides the interfaces described in Sect. 2.3 (Fig. 3) and implements multi-version concurrency control.

Each transaction is modeled as a sequence of operations: *i*) fetch a data item; *ii*) do some processing; *iii*) write back a data item. Upon receiving a transaction request each operation is scheduled to execute on the corresponding resource. The processing time of each operation is previously obtained by profiling a real database server (Sect. 3.2).

A database client is attached to a database server and produces a stream of transaction requests. After each request is issued, the client blocks until the server replies, thus modeling a single threaded client process. After receiving a reply, the client is then paused for some amount of time (thinking time) before issuing the next transaction request.

To determine the read-set and write-set of a transaction's execution, the database is modeled as a set of histograms [5]. The transactions' statements are executed against this model and the read-set, write-set and write-values are extracted to build the transaction model that is injected into the database server. In our case, this modeling is rather straightforward as the database is very well defined by the TPC-C [23] workload that we use for all tests. Moreover, as all the transactions specified by TPC-C can be reduced to SPJ queries, the read-set extraction is quite simple.

3.2   Test Parameters

Each database request is generated according to the TPC-C benchmark [23]. TPC-C is the industry standard on-line transaction processing benchmark. It mimics a wholesale supplier with a number of geographically distributed sales districts and associated warehouses. TPC-C specifies a precise set of relations (Warehouse, District, Customer, Item, Stock, Orders, OrderLine, NewOrder and History) and the size of the database as a function of the number of desired clients. The benchmark determines 10 clients per warehouse and, as an example, for 2000 clients, the database contains around 10^9 tuples, each ranging from 8 to 655 bytes. The traffic is a mixture of read-only and update intensive transactions. A client can request transactions of five different types:

NewOrder, adds a new order into the system (with 44% of the occurrences); *Payment*, updates the customer's balance, district and warehouse statistics (44%); *OrderStatus*, returns a given customer latest order (4%); *Delivery*, records the delivery of products (4%); *StockLevel*, determines the number of recently sold items that have a stock level below a specified threshold (4%). The *NewOrder*, *Payment* and *Delivery* are update transactions while the others are read-only.

The database model has been configured using the transactions' processing time of a profiled version of PostgreSQL 7.4.6 under the TPC-C workload. From the TPC-C benchmark we only use the specified workload, the constraints on throughput, performance, screen load and background execution of transactions are not taken into account.

We consider a LAN and a WAN scenarios, both with 9 replicas. In the LAN configuration the replicas are connected by a network with a bandwidth of 1Gbps and a latency of 120μs. The WAN configuration consists of 3 LANs (with 1Gbps and 120μs as before) each with 3 replicas, connected by a network with a bandwidth of 100Mbps and a latency of 60ms. Each replica corresponds to a dual processor AMD Opteron at 2.4GHz with 4GB of memory, running the Linux Fedora Core 3 Distribution with kernel version 2.6.10. For storage we used a fiber-channel attached box with 4, 36GB SCSI disks in a RAID-5 configuration and the Ext3 file system.

For all the experiments that follow, we varied the total of clients from 270 to 3960 and distributed them evenly among the replicas.

4 Experimental Results

4.1 Simple Configuration

The first scenario evaluates the conservative and the DBSM approaches without exploiting any application specific details and thus in a configuration that can easily be automated. In the conservative approach, we use the simple definition of a conflict class for each table, which can actually be easily extracted from the SQL code. The resulting conflict classes and conflict relations among transactions types are shown in the "Serializable" column of Table 1. Regarding the DBSM, we need to pay special attention to read-set sizes since the propagation of large read-sets may be impractical. An immediate workaround to this problem is to set a limit for the read-set size over which the whole table is used. In the TPC-C, this results in transactions of type *Delivery* always being marked as reading the entire *OrderLine* table. All others access only a small number of items.

Figure 4 presents performance measurements in the LAN scenario. It can be observed in Fig. 4(a) that the DBSM protocol with optimistic execution apparently scales much better to a large number of clients than the conservative protocol. As shown by Fig. 4(b) the bottleneck in the conservative protocol translates in very large queueing latencies.

However, as seen in Fig. 4(c), the good throughput of the DBSM is achieved at the expense of a number of aborted transactions. This is especially worrisome since the 4% of transactions being aborted overall are in fact all *Delivery* transactions as shown in Fig. 4(d). Therefore, even if such transactions can be resubmitted, there is a very low probability of ever being executed. These results show that neither of the approaches scale to a large number of clients with an OLTP load, even with plenty of resources in a LAN.

Table 1. Definition of conflict classes for each transaction type in TPC-C

Class./Trans	Serializable			Snapshot Isolation Level		
	New Order	Payment	Delivery	New Order	Payment	Delivery
Warehouse	X	X			X	
District	X	X		X	X	
Customer	X	X	X		X	X
Item	X					
Stock	X			X		
Orders	X		X	X		X
OrderLine	X		X	X		X
NewOrder	X		X	X		X
History		X			X	

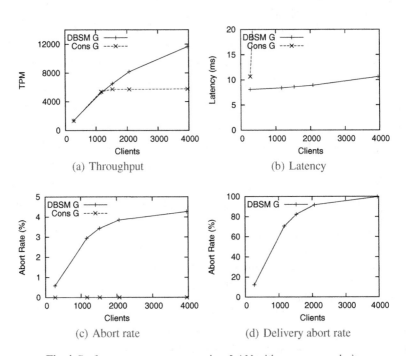

(a) Throughput

(b) Latency

(c) Abort rate

(d) Delivery abort rate

Fig. 4. Performance measurements in a LAN with coarse granularity

4.2 Fine Granularity

To reduce the number of conflicts, we resort to a finer granularity when defining conflict classes for the conservative approach and the read-set extraction in the DBSM. Fine grained conflict classes are obtained by taking advantage of the fact that all tables except *Item* have references to the *Warehouse* table and that clients connected to the same node have high locality regarding a specific subset of warehouses.

Although this may seem to easily translate in a definition of conflict classes for the conservative protocol, in practice it is not possible because transactions *Payment* and

NewOrder, which account for a large majority of traffic, may access multiple warehouses. Despite the suitability of this assumption to the TPC-C workload, it must not be generalized since most of the time one cannot be certain of which subset of a table a transaction will access, rendering the approach impractical.

In the optimistic protocol, one uses the same observation to avoid a huge read set without escalating to table level by using only the warehouse attribute and then encoding it as part of the table identifier.

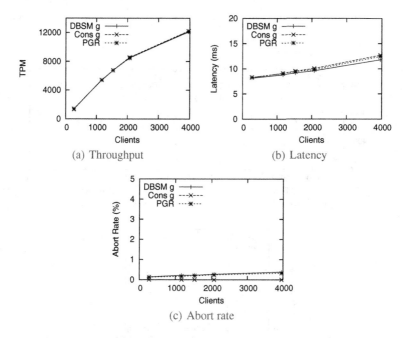

(a) Throughput

(b) Latency

(c) Abort rate

Fig. 5. Performance measurements in a LAN with fine granularity

We compare also these optimizations with the PGR protocol which can use the exact read-set by centralizing certification of each transaction. The results are presented in Fig. 5. It can be observed that all approaches produce approximate results with minimal differences in latency and abort rate. Network usage is also very close, showing that the overhead incurred by the DBSM when sending the read-set is offset by requiring only a single communication step. These results show that with an appropriate granularity, all these group communication based protocols are equally appropriate for an OLTP load in a cluster.

4.3 Snapshot Isolation

An alternative approach to avoid synchronization conflicts is to relax the correctness criterion to snapshot isolation [2] which only considers write-write conflicts.

In the DBSM approach, all the concerns previously discussed about the size of the read-set are avoided. As Fig. 6 shows, it turns out that this alternative has also a benign impact on the performance of the DBSM approach, reducing the number of aborted transactions. Moreover, this is a very appealing alternative, as it avoids all configuration issues. As explained in Sect. 2.2, under snapshot isolation the DBSM and PGR protocols become the same.

Unlike the DBSM, the conservative approach does not benefit from the snapshot isolation criterion, exhibiting the same latency as before. In the "Snapshot Isolation Level" column of Table 1 the new conflict relations among the transactions are depicted. Regardless of their type, all update transactions still conflict and thus have to be sequentially executed.

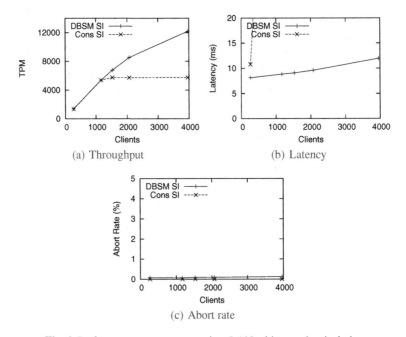

(a) Throughput

(b) Latency

(c) Abort rate

Fig. 6. Performance measurements in a LAN with snapshot isolation

4.4 Wide Area

Finally, we are interested in observing how the proposed approaches scale also to interconnected clusters in WAN. The best performers in the previous scenarios were chosen and their performance in this environment is presented Fig. 7. Although Fig. 7(a) shows that throughput scales equally well, Fig. 7(b) shows that the additional communication step, incurred by PGR, when centralizing certification results in a large increase in latency. This has also an impact in the overall abort rate in Fig. 7(c), which is higher than with other optimistic approaches. Note however that, in contrast with the results of Fig. 4, Fig. 7(d) shows that no single transaction type exhibits high abort rates, hence, if one chooses to resubmit the aborted transactions there is a high probability of a successful execution.

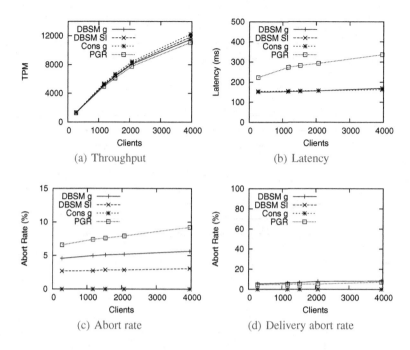

Fig. 7. Performance measurements in a WAN

4.5 Discussion

The key issue in obtaining close to linear scalability of a distributed system is reducing synchronization overhead. In practice, one can measure this overhead by the time the computation in a node is suspended waiting for interaction with remote nodes. In a traditional protocol based on distributed locking, this can potentially be very large, if a node has to wait that all other nodes enter and leave a critical section plus the time it takes to pass the authorization around.

In contrast, when using active replication [19,8], the only overhead is encapsulated in the total order multicast protocol and no additional synchronization is required. Ideally, a database replication protocol based on total order multicast would be able to achieve the same goal. We now examine in turn each of the protocols to determine how this goal is achieved.

Figure 8 depicts the conservative and optimistic protocols handling the execution of two concurrent non-conflicting transactions. In the CONS protocol (Fig. 8(a)), once the transactions are ordered all steps of the protocol are executed concurrently therefore corresponding to the desired behavior.

Regarding the optimistic approaches, we can see that in the DBSM (Fig. 8(b)) the transactions' execution can always be carried in parallel while the certification procedure needs to be done sequentially. Once the certification is finished, since the transactions do not conflict, the updates may be incorporated concurrently. The DBSM therefore incurs in the certification procedure overhead. However, the certification execution time is usually negligible though.

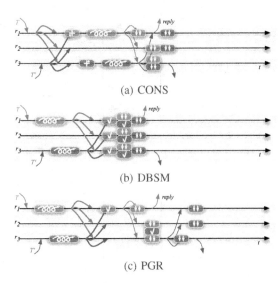

(a) CONS

(b) DBSM

(c) PGR

Fig. 8. Handling concurrent transactions

In contrast, the PGR protocol is penalized by the supplemental reliable multicast. Although the transactions' execution can be done in parallel too, the certification of T' (ordered after T) can only be done once r_3 knows the outcome of T. That is, the latency of the reliable multicast of T is incorporated in the response time of T'. This problem can further suffer a cascading effect caused by the expected system parallelism.

5   Conclusions

Database replication protocols based on group communication have been previously evaluated with a variety of implementation or simulation techniques and also a variety of, often non-representative, loads or system models. When an industry standard database benchmark is used, it is often TPC-W, which provides a read-intensive load which does not stress synchronization mechanisms. This makes it difficult to compare their relative trade-offs and performance regarding non-replicated databases.

In contrast, in this paper, we use the realistic write-intensive OLTP load from the TPC-C benchmark and in Sect. 4 we show that high performance and close to linear scalability can be achieved with several configurations. In detail, we show that when snapshot isolation suffices for the application requirements, as happens with TPC-C itself, the DBSM-SI protocol is the best option, requiring little effort to configure and offering excellent performance in LAN and WAN. When serializability is required, there are two possible options. When adequately fine grained conflict classes can be defined, predicted beforehand for each transaction, and the source modified to convey them, a conservative protocol provides excellent performance without introducing transaction aborts. If adequately fine-grained conflict classes cannot be predicted or the source modified to tag transactions or cope with snapshot isolation, as happens when supporting large third party legacy applications, the DBSM protocol provides the same performance when given an adequate definition of read-set granularity. Notice that this can be achieved by a database administrator with no modification of sources and with no impact on correctness, which provides maximum flexibility and safety.

In short, group communication based database replication protocols provide a spectrum of configurability, generality and performance trade-offs that fit the most demanding applications. The wide availability of such protocols therefore demands improved database interfaces that efficiently provide the functionality identified in Sect. 2 of this paper.

References

1. D. Agrawal, A. El Abbadi, and R. C. Steinke. Epidemic algorithms in replicated databases (extended abstract). In *Proc. ACM Symp. Principles of Database Systems (PODS)*, 1997.
2. H. Berenson, P. Bernstein, J. Gray, J. Melton, E. O'Neil, and P. O'Neil. A critique of ANSI SQL isolation levels, 1995.
3. P. Bernstein, V. Hadzilacos, and N. Goodman. *Concurrency Control and Recovery in Database Systems*. Addison-Wesley, 1987.
4. N. Budhiraja, K. Marzullo, F. Schneider, and S. Toueg. The primary-backup approach. In S. Mullender, editor, *Distributed Systems*, chapter 8. Addison Wesley, 1993.
5. A. Correia, A. Menezes, and R. Oliveira. Off-line test automation for database replication based on group communication. Technical report, Universidade do Minho, 2005.
6. J. Cowie. *Scalable Simulation Framework API Reference Manual*, March 1999.
7. J. Cowie, H. Liu, J. Liu, D. Nicol, and Andy Ogielski. Towards realistic million-node internet simulation. In *Proc. Int'l Conf. Parallel and Distributed Processing Techniques and Applications (PDPTA'99)*, 1999.
8. R. Guerraoui and A. Schiper. Software-based replication for fault tolerance. *IEEE Computer*, 30(4), April 1997.
9. S. Guo, W. Sun, and M. Weiss. Solving Satisfiability and Implication Problems in Database Systems. *ACM Transactions on Database Systems (TODS)*, 1996.
10. V. Hadzilacos and S. Toueg. A modular approach to fault-tolerant broadcasts and related problems. Technical Report TR94-1425, Cornell Univ., Computer Science Dept., May 1994.
11. J. Holliday, D. Agrawal, and A. El Abbadi. The performance of database replication with group multicast. In *Proc. IEEE Int'l Symp. Fault-Tolerant Computing Systems (FTCS)*, 1999.
12. B. Kemme and G. Alonso. A suite of database replication protocols based on communication primitives. In *Proc. IEEE Int'l Conf. Distributed Computing Systems (ICDCS)*, 1998.
13. B. Kemme and G. Alonso. Don't be lazy, be consistent: Postgres-r, a new way to implement database replication. In *VLDB '00: Proceedings of the 26th International Conference on Very Large Data Bases*, pages 134–143, San Francisco, CA, USA, 2000. Morgan Kaufmann Publishers Inc.
14. B. Kemme, F. Pedone, G. Alonso, and A. Schiper. Processing transactions over optimistic atomic broadcast protocols. In *Proc. IEEE Int'l Conf. Distributed Computing Systems (ICDCS)*, 1999.
15. M. Patiño-Martínez, R. Jiménez-Peris, B. Kemme, and G. Alonso. Scalable replication in database clusters. In *DISC'00: Proceedings of the 14th International Conference on Distributed Computing*, pages 315–329, London, UK, 2000. Springer-Verlag.
16. F. Pedone. *The Database State Machine and Group Communication Issues*. PhD thesis, Département d'Informatique, École Polytechnique Fédérale de Lausanne, 1999.
17. F. Pedone, R. Guerraoui, and A. Schiper. The database state machine approach. *J. Distributed and Parallel Databases and Technology*, 2003.

18. M. Pettersson. Linux performance counters. http://user.it.uu.se/ mikpe/linux/perfctr/, 2004.

19. F. Schneider. Replication management using the state-machine approach. In S. Mullender, editor, *Distributed Systems*, chapter 7. Addison Wesley, 1993.

20. A. Sousa, F. Pedone, R. Oliveira, and F. Moura. Partial replication in the database state machine. In *IEEE Int'l Symp. Networking Computing and Applications*, 2001.

21. A. Sousa, J. Pereira, F. Moura, and R. Oliveira. Optimistic total order in wide area networks. In *Proc. IEEE Int'l Symp. Reliable Distributed Systems (SRDS)*, 2002.

22. A. Sousa, J. Pereira, L. Soares, A. Correia Jr., L. Rocha, R. Oliveira, and F. Moura. Testing the dependability and performance of group communication based database replication protocols. In *IEEE Intl. Conf. on Dependable Systems and Networks - Performance and Dependability Symposium (DSN-PDS'2005)*, 2005. to appear.

23. Transaction Processing Performance Council (TPC). TPC Benchmark™ C standard specification revision 5.0, February 2001.

24. S. Wu and B. Kemme. Postgres-r(si): Combining replica control with concurrency control based on snapshot isolation. In *Proc. of the IEEE Int. Conf. on Data Engineering (ICDE)*, pages 422–433, April 2005.

Third Workshop on Theses and Dissertations on Dependable Computing

Avelino Zorzo[1], Ingrid Jansch-Pôrto[2], and Fabíola Gonçalves Pereira Greve[3]

[1] Pontifícia Universidade Católica do Rio Grande do Sul,
Computing Science Department
zorzo@inf.pucrs.br
[2] Institute of Informatics,
Federal University of Rio Grande do Sul
ingrid@inf.ufrgs.br
[3] Department of Computer Science,
Federal University of Bahia
fabiola@ufba.br

The Workshop on Theses and Dissertations on Dependable Computing is a student forum for bringing together graduate students that research on topics related to dependable computing. The aim of this meeting is to present and discuss the proposed contribution, preliminary results and possible directions for their research. The previous editions of this Workshop were held in Florianópolis in conjunction with the Brazilian Symposium on Fault Tolerance (SCTF 2001), and in 2003 in São Paulo with the Latin-American Symposium on Dependable Computing (LADC 2003).

C.A. Maziero et al. (Eds.): LADC 2005, LNCS 3747, p. 261, 2005.
© Springer-Verlag Berlin Heidelberg 2005

Latin-American Workshop on Dependable Automation Systems

Herman Augusto Lepikson[1] and Leandro Buss Becker[2]

[1] Department of Mechanical Engineering,
Federal University of Bahia
herman@ufba.br
[2] Department of Automation and Systems,
Federal University of Santa Catarina
lbecker@das.ufsc.br

Automation systems play an important role in the economy of most industrialized countries. One prominent feature of such systems relates to dependability, as unexpected crashes can both put human-life in danger and cause massive money looses. The first Latin-American Workshop on Dependable Automation Systems (WDAS) aims to provide an opportunity for researchers and industrial partners to discuss problems related to the development of safe-critical automation systems.

Topics covered in our one-day workshop on Dependable Automation Systems include the automatic detection of software failures, fault analysis and diagnosis for dependable automation systems, reliability in real-time automation systems, dependable real-time control/coordination systems, and safety communication over wireless networks. The program is composed of regular papers, invited papers, a key note speech, and a panel.

We expect to promote a pleasant environment for technical discussions, bringing together researchers and practitioners to share research results, practical experiences, and advances in (or impediments to) the application of dependability concepts for building automation systems. We encourage participation by professionals with diverse backgrounds who can contribute to advancing the technology and reflecting the latest trends and who can foster discussing the implications.

Finally, we want to thank everybody who contributed to this workshop, first of all, the colleagues which provided the technical contributions to the workshop program. Our special thank also goes to the members of the program committee who helped to shape the workshop with their recommendations. Everything is set for a stimulating and hopefully highly interactive event. Thank you for your interest and enjoy the technical program and discussions.

C.A. Maziero et al. (Eds.): LADC 2005, LNCS 3747, p. 262, 2005.
© Springer-Verlag Berlin Heidelberg 2005

Software Architectures for Dependable Systems

Rogério de Lemos[1] and Paulo Asterio de Castro Guerra[2]

[1] University of Kent, Computing Laboratory,
Canterbury, Kent CT2 7NF, UK
r.delemos@kent.ac.uk
http://www.cs.kent.ac.uk/people/staff/rdl/
[2] R. José Santoro, 17, 37480-000 - Lambari, MG, Brazil
asterio@acm.org

Abstract. Although there is a large body of research in dependability, architectural level reasoning about dependability is only just emerging as an important theme in software development. This is due to the fact that dependability concerns are often left until too late in the process of development. In addition, the complexity of emerging applications and the trend of building trustworthy systems from existing untrustworthy components are urging dependability concerns to be considered at the architectural level.

1 Motivation

The structure of a system is what enables it to generate its intended behaviour, from the behaviour of its components. The architecture of a software system is an abstraction of the actual structure of that system. The identification of the system structure early in its development process allows abstracting away from details of the system, thus assisting the understanding of broader system concerns [2]. One of the benefits of a well-structured system is to avoid overly complex relationships between its components, which in turn should lead to a more dependable system. Dependability can be defined as the ability of a system to deliver service that can justifiably be trusted [1].

Reasoning about dependability at the architectural level has lately grown in importance because of the complexity of emerging applications, and the trend of building them through the integration of pre-existing software components. This component-based trend requires trustworthy systems to emerge from the integration of untrustworthy components, whose actual implementations may even not be known in advance. For instance, the deployment of a new version of an "off-the-shelf" (OTS) component in the environment of a trustworthy component-based system should not represent a risk to the dependability of that system, even when the new OTS version introduces new faults in that same system. As a result, these new applications demand for dependability concerns to be considered at the architectural level, rather than late in the development process. From the perspective of software engineering, which strives to build software systems that are free of faults, the architectural consideration of dependability compels the acceptance of residual and unanticipated faults, rather than relying only in

C.A. Maziero et al. (Eds.): LADC 2005, LNCS 3747, pp. 263–264, 2005.
© Springer-Verlag Berlin Heidelberg 2005

their avoidance. Thus the need for novel notations, methods and techniques that provides the necessary support for reasoning about faults at the architectural level [3]. For example, notations should be able to represent non-functional properties and failure assumptions, and techniques should be able to extract from the architectural representations the information that is relevant for evaluating the system architecture from a certain perspective.

2   Outline

The tutorial presents the current academic research by addressing the following main topics:

- **Introduction to software architectures and dependability**
  - Basic concepts in software architectures: architectural styles, and architectural description languages (ADLs).
  - Basic concepts in dependability: threats, attributes and means.
- **Architecting for dependability**
  - *Rigorous design*: architectural-based software development, ADLs for dependability, UML as an architectural description language, and wrappers and protectors.
  - *Verification and validation*: architectural model checking.
  - *Fault tolerance*:exception handling in software architectures, idealized fault tolerant architectural elements, N-version programming and recovery blocks at the architectural level, and architectural reconfiguration.
  - *System evaluation*: qualitative analysis, Architecture Analysis Tradeoff Method (ATAM), quantitative analysis, and stochastic techniques.
- **Future trends**

References

1. A. Avizienis, J.-C. Laprie, B. Randell, C. Landwehr. "Basic Concepts and Taxonomy of Dependable and Secure Computing". *IEEE Transactions on Dependable and Secure Computing* **1(1)**. January-March 2004. pp. 11-33.
2. P. Clements, et al. *Documenting Software Architectures: Views and Beyon.* Addison-Wesley. 2003.
3. C. Gacek, R. de Lemos. "Architectural Description of Dependable Software Systems". *Structuring Computer-based Systems for Depedanbility.* C. Jones, C. Gacek, D. Besnard (Eds.). Springer. 2005.

Fault-Tolerant Techniques for Concurrent Objects

Rachid Guerraoui[1] and Michel Raynal[2]

[1] Ecole Polytechnique Fédérale de Lausanne,
Département d'Informatique, 1015 Lausanne, Switzerland
`Rachid.Guerraoui@epfl.ch`
[2] IRISA, Campus de Beaulieu, Université de Rennes,
Avenue du Général Leclerc, 35042 Rennes Cedex, France
`raynal@irisa.fr`

Devising wait-free resilient implementations of concurrent objects from fault-prone base objects is a fundamental challenge of computer science. Wait-free means that any process that invokes an operation eventually receives a reply after executing a finite number of its own steps, even if other processes are arbitrarily slow or even failed. Resilience means that the implementation of the concurrent object behaves correctly despite the failure of up to t base objects (t being a threshold parameter a priori defined). The tutorial surveys different techniques to build wait-free resilient implementations of concurrent objects. Three complementary classes of techniques are presented: (1) fault-tolerance "by replication", (2) fault-tolerance "by diversity", and (3) fault-tolerance "by oracle", respectively. The first is the well-known redundancy technique and its applicability depends on the kinds of faults that the objects can suffer. The second consists in combining the base objects with objects of other types (type refers here to a programming language notion: the type has to be powerful enough to allow implementing resilient objects). This technique basically relies on the universality of consensus objects. The third technique relies on the information we can obtain about the operational status of the processes.

The aim of the tutorial is to make people familiar with practical and theoretical fault-tolerance techniques and concepts to build resilient concurrent objects. To illustrate the techniques, the tutorial uses algorithms from the literature or devises new algorithms. A simple framework to derive a family of consensus algorithms tolerating process crash failures and asynchronous periods, will be presented. This framework is based on two independent abstractions, Alpha and Omega, that cleanly address orthogonal issues: Alpha is devoted to consensus safety, while Omega is devoted to consensus liveness. Implementations of the Alpha abstraction in shared memory, storage area network, message passing and active disk systems will be presented, leading to directly derive consensus algorithms suited to these communication media. (Interestingly, the algorithms derived from the framework can be viewed as variants of the Paxos seminal consensus algorithm of Lamport. In this sense, this part of the tutorial can be seen as guided visit to variants of Paxos algorithms that have appeared recently in the literature.)

C.A. Maziero et al. (Eds.): LADC 2005, LNCS 3747, p. 265, 2005.

Agreement Protocols in Environments with Temporal Uncertainties

Fabíola Gonçalves Pereira Greve

Department of Computer Science. Federal University of Bahia,
Campus de Ondina, 40170-110 Bahia, Brazil
fabiola@ufba.br

Agreement protocols are fundamental for the design of dependable systems. They ensure consistent cooperation among distributed entities, helping both to keep the continuity of services in spite of failures and to enhance performance. Consensus is the greatest common denominator among all agreement problems. It allows a set of processes to agree on a common output value. Theoretical advances have been reached, thanks to the consensus problem solutions through the use of unreliable failure detectors, which have been proved to be essential in solving many other agreement problems in environments with temporal uncertainties. Such advances have been exploited in order to (i) find efficient solutions to agreement problems, (ii) identify minimal synchronous conditions for their solution and (iii) characterize more precisely their behavior (blocking or progression) in presence of network disturbs. From a software engineering view point, consensus-based protocols give rise to simple and modular solutions. Basic components (*consensus, reliable broadcast, atomic broadcast, failure detector, etc.*) are identified in order to construct richer ones (*group membership, view synchrony, atomic commit, etc.*). These components are in turn the fundamental pieces of middleware for reliable distributed programming.

This tutorial presents a survey of the latest advances in solving agreement in environments with temporal uncertainties. Firstly, recent theoretical results regarding the solutions of agreement problems as well as their algorithms are presented. Afterwards, it is shown how these algorithms are combined to build services for fault-tolerant middleware. These are group and replication management systems. Finally, through an example of task allocation in a computational grid, it is shown how these protocols and middleware could be used in both the design and the implementation of dependable applications.

C.A. Maziero et al. (Eds.): LADC 2005, LNCS 3747, p. 266, 2005.
© Springer-Verlag Berlin Heidelberg 2005

Author Index

Lecture Notes in Computer Science

For information about Vols. 1–3658

please contact your bookseller or Springer

Vol. 3702: B. Beckert (Ed.), Automated Reasoning with Analytic Tableaux and Related Methods. XIII, 343 pages. 2005. (Subseries LNAI).

Vol. 3701: M. Coppo, E. Lodi, G. M. Pinna (Eds.), Theoretical Computer Science. XI, 411 pages. 2005.

Vol. 3699: C.S. Calude, M.J. Dinneen, G. Păun, M. J. Pérez-Jiménez, G. Rozenberg (Eds.), Unconventional Computation. XI, 267 pages. 2005.

Vol. 3698: U. Furbach (Ed.), KI 2005: Advances in Artificial Intelligence. XIII, 409 pages. 2005. (Subseries LNAI).

Vol. 3697: W. Duch, J. Kacprzyk, E. Oja, S. Zadrożny (Eds.), Artificial Neural Networks: Formal Models and Their Applications – ICANN 2005, Part II. XXXII, 1045 pages. 2005.

Vol. 3696: W. Duch, J. Kacprzyk, E. Oja, S. Zadrożny (Eds.), Artificial Neural Networks: Biological Inspirations – ICANN 2005, Part I. XXXI, 703 pages. 2005.

Vol. 3695: M.R. Berthold, R. Glen, K. Diederichs, O. Kohlbacher, I. Fischer (Eds.), Computational Life Sciences. XI, 277 pages. 2005. (Subseries LNBI).

Vol. 3694: M. Malek, E. Nett, N. Suri (Eds.), Service Availability. VIII, 213 pages. 2005.

Vol. 3693: A.G. Cohn, D.M. Mark (Eds.), Spatial Information Theory. XII, 493 pages. 2005.

Vol. 3692: R. Casadio, G. Myers (Eds.), Algorithms in Bioinformatics. X, 436 pages. 2005. (Subseries LNBI).

Vol. 3691: A. Gagalowicz, W. Philips (Eds.), Computer Analysis of Images and Patterns. XIX, 865 pages. 2005.

Vol. 3690: M. Pěchouček, P. Petta, L.Z. Varga (Eds.), Multi-Agent Systems and Applications IV. XVII, 667 pages. 2005. (Subseries LNAI).

Vol. 3689: G.G. Lee, A. Yamada, H. Meng, S.H. Myaeng (Eds.), Information Retrieval Technology. XVII, 735 pages. 2005.

Vol. 3688: R. Winther, B.A. Gran, G. Dahll (Eds.), Computer Safety, Reliability, and Security. XI, 405 pages. 2005.

Vol. 3687: S. Singh, M. Singh, C. Apte, P. Perner (Eds.), Pattern Recognition and Image Analysis, Part II. XXV, 809 pages. 2005.

Vol. 3686: S. Singh, M. Singh, C. Apte, P. Perner (Eds.), Pattern Recognition and Data Mining, Part I. XXVI, 689 pages. 2005.

Vol. 3685: V. Gorodetsky, I. Kotenko, V. Skormin (Eds.), Computer Network Security. XIV, 480 pages. 2005.

Vol. 3684: R. Khosla, R.J. Howlett, L.C. Jain (Eds.), Knowledge-Based Intelligent Information and Engineering Systems, Part IV. LXXIX, 933 pages. 2005. (Subseries LNAI).

Vol. 3683: R. Khosla, R.J. Howlett, L.C. Jain (Eds.), Knowledge-Based Intelligent Information and Engineering Systems, Part III. LXXX, 1397 pages. 2005. (Subseries LNAI).

Vol. 3682: R. Khosla, R.J. Howlett, L.C. Jain (Eds.), Knowledge-Based Intelligent Information and Engineering Systems, Part II. LXXIX, 1371 pages. 2005. (Subseries LNAI).

Vol. 3681: R. Khosla, R.J. Howlett, L.C. Jain (Eds.), Knowledge-Based Intelligent Information and Engineering Systems, Part I. LXXX, 1319 pages. 2005. (Subseries LNAI).

Vol. 3680: C. Priami, A. Zelikovsky (Eds.), Transactions on Computational Systems Biology II. IX, 153 pages. 2005. (Subseries LNBI).

Vol. 3679: S.d.C. di Vimercati, P. Syverson, D. Gollmann (Eds.), Computer Security – ESORICS 2005. XI, 509 pages. 2005.

Vol. 3678: A. McLysaght, D.H. Huson (Eds.), Comparative Genomics. VIII. 167 pages. 2005. (Subseries LNBI).

Vol. 3677: J. Dittmann, S. Katzenbeisser, A. Uhl (Eds.), Communications and Multimedia Security. XIII, 360 pages. 2005.

Vol. 3676: R. Glück, M. Lowry (Eds.), Generative Programming and Component Engineering. XI, 448 pages. 2005.

Vol. 3675: Y. Luo (Ed.), Cooperative Design. Visualization, and Engineering. XI, 264 pages. 2005.

Vol. 3674: W. Jonker, M. Petković (Eds.), Secure Data Management. X, 241 pages. 2005.

Vol. 3673: S. Bandini, S. Manzoni (Eds.), AI*IA 2005: Advances in Artificial Intelligence. XIV, 614 pages. 2005. (Subseries LNAI).

Vol. 3672: C. Hankin, I. Siveroni (Eds.), Static Analysis. X, 369 pages. 2005.

Vol. 3671: S. Bressan, S. Ceri, E. Hunt, Z.G. Ives, Z. Bellahsène, M. Rys, R. Unland (Eds.), Database and XML Technologies. X, 239 pages. 2005.

Vol. 3670: M. Bravetti, L. Kloul, G. Zavattaro (Eds.), Formal Techniques for Computer Systems and Business Processes. XIII, 349 pages. 2005.

Vol. 3669: G.S. Brodal, S. Leonardi (Eds.), Algorithms – ESA 2005. XVIII, 901 pages. 2005.

Vol. 3668: M. Gabbrielli, G. Gupta (Eds.), Logic Programming. XIV, 454 pages. 2005.

Vol. 3666: B.D. Martino, D. Kranzlmüller, J. Dongarra (Eds.), Recent Advances in Parallel Virtual Machine and Message Passing Interface. XVII, 546 pages. 2005.

Vol. 3665: K. S. Candan, A. Celentano (Eds.), Advances in Multimedia Information Systems. X, 221 pages. 2005.

Vol. 3664: C. Türker, M. Agosti, H.-J. Schek (Eds.), Peer-to-Peer, Grid, and Service-Orientation in Digital Library Architectures. X, 261 pages. 2005.

Vol. 3663: W.G. Kropatsch, R. Sablatnig, A. Hanbury (Eds.), Pattern Recognition. XIV, 512 pages. 2005.

Vol. 3662: C. Baral, G. Greco, N. Leone, G. Terracina (Eds.), Logic Programming and Nonmonotonic Reasoning. XIII, 454 pages. 2005. (Subseries LNAI).

Vol. 3661: T. Panayiotopoulos, J. Gratch, R. Aylett, D. Ballin, P. Olivier, T. Rist (Eds.), Intelligent Virtual Agents. XIII, 506 pages. 2005. (Subseries LNAI).

Vol. 3660: M. Beigl, S. Intille, J. Rekimoto, H. Tokuda (Eds.), UbiComp 2005: Ubiquitous Computing. XVII, 394 pages. 2005.

Vol. 3659: J.R. Rao, B. Sunar (Eds.), Cryptographic Hardware and Embedded Systems – CHES 2005. XIV, 458 pages. 2005.